STP 1326

Oxidative Behavior of Materials by Thermal Analytical Techniques

Alan T. Riga and Gerald H. Patterson, editors

ASTM Publication Code Number (PCN):
04-013260-50

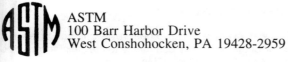

ASTM
100 Barr Harbor Drive
West Conshohocken, PA 19428-2959

Printed in the U.S.A.

Library of Congress Cataloging-in-Publication Data

Oxidative behavior of materials by thermal analytical techniques /
   Alan T. Riga and Gerald H. Patterson, editors.
      p. cm.—(STP : 1326)
     "ASTM publication code number (PCN): 04-013260-50."
     Includes bibliographical references and index.
     ISBN 0-8031-2483-X
     1. Thermal analysis—Congresses. 2. Oxidation—Congresses.
   I. Riga, Alan T. II. Patterson, Gerald H., 1937–   . III. Series:
ASTM special technical publication ; 1326.
QD117.T4095 1997
620.1'173—DC21   97-39992
                                                  CIP

Photocopy Rights

Authorization to photocopy items for internal, personal, or educational classroom use, or the internal, personal, or educational classroom use of specific clients, is granted by the American Society for Testing and Materials (ASTM) provided that the appropriate fee is paid to the Copyright Clearance Center, 222 Rosewood Dr., Danvers, MA 01923; Tel: (508) 750-8400, online: http://www.copyright.com/.

Peer Review Policy

Each paper published in this volume was evaluated by two peer reviewers and at least one of the editors. The authors addressed all of the reviewers' comments to the satisfaction of both the technical editor(s) and the ASTM Committee on Publications.

To make technical information available as quickly as possible, the peer-reviewed papers in this publication were prepared "camera-ready" as submitted by the authors.

The quality of the papers in this publication reflects not only the obvious efforts of the authors and the technical editor(s), but also the work of these peer reviewers. The ASTM Committee on Publications acknowledges with appreciation their dedication and contribution to time and effort on behalf of ASTM.

Printed in Philadelphia, PA
October, 1997

Foreword

This publication, *Oxidative Behavior of Materials by Thermal Analytical Techniques,* contains papers presented at the symposium of the same name, held in New Orleans, Louisiana, on 21-22 Nov. 1996. The symposium was sponsored by ASTM Committee E37 on Thermal Methods. Alan Riga and Gerald Patterson served as chairpersons of the symposium and are editors of the resulting publication.

Contents

Overview

This Special Technical Publication represents a compliation of presentations from an international symposium addressing the Oxidative Behavior of Materials by Thermal Analytical Techniques which was held 20-21 Nov. 1996 in New Orleans, Louisiana. The symposium and this subsequent publication examine the new thermal analytical techniques describing the physical properties and oxidative degradative behavior of polymers, lubricants, and petrochemicals. Historical reviews, oxidation mechanisms, new test methods, unique techniques, robotic methods, new reference standards, and bias considerations form the basis of this publication.

It was generally agreed, but certainly highlighted by Roger Blaine, TAI, keynote speaker, that the factors affecting the oxidation induction time (OIT) were: the isothermal temperature; pan type, metallurgy, and shape; and pressure and oxygen flow rate. Polymer oxidation was emphasized in a number of papers. Professor Joseph Perez, Penn State University, keynote speaker, focused on the measurement of oxidation in formulated passenger car and diesel engine oils. Professor Perez discussed the evolution of oxidation systems from the complex Dornte type systems of the 1940s to the current use of microreactors, for example, DSC, TGA, and Klaus Penn State Microreactor (PSMO). He stressed the variables such as: metal surfaces that affect the rate of oxidation or thermal decomposition, temperature effects on the related thermal and oxidative processes, effects of volatility on vapor phase versus liquid phase oxidation, oxygen diffusion rate limitations, and additive effectiveness. Some of the differences in the thermal and oxidative behavior in bulk systems (engine tests) and microsystems (DSC, TGA, PSMO) were discussed.

Alan Riga, Lubrizol, reviewed the recently approved and soon to be published Standard Test Method for Determing OIT of Hydrocarbons by DSC/PDSC. An interlaboratory study using this DSC/PDSC method was reported by Blaine and Riga, with ASTM Reference C, a diluted fully formulated engine oil from Lubrizol and ASTM Reference D, a polyethylene film from TAI.

A successful robotic DSC evaluation by M. Kelsey, Mettler-Toledo, clearly differentiated References C and D at 195°C and one atmosphere of oxygen. This abbreviated OIT method without heating/cooling (from room temperature) or gas switching reduces the experimental time, avoids a nitrogen purge, and gives a slightly better reproducibility. The abbreviated OIT method suggests a possible revision to the DSC ASTM E37 test protocol. A question raised by a number of participants was "Is it possible that some PDSC oxidative testing can be replaced with robotic DSC?"

A modified Arrehenius model was used to predict OIT of polymers. An adaptation of this model for evaluation of the oxidative stability of oils was discussed.

A new scanning DTA/TGA oxidation test was presented. This DTA method is based on air oxidation and defines an oxidation temperature. A good correlation was observed between the isothermal PDSC OIT in oxygen of eight readily available olefin reference polymers and the oxidation temperature. The DTA air oxidation method as well as the PDSC oxygen method use the melting temperatures and heat of fusion of polyethylenes and polypropylenes to verify the temperature and heat calibrations.

A unique approach to the oxidation process was presented by Rick Seyler (Kodak). This paper considered oxidation as a negative factor in the determination of vapor pressure by

DSC in nitrogen. The observed exotherms in DSC measurements were associated with partial oxidation of the chemical specimen from residual air in the DSC specimen container. An application of this method is to study vapor phase oxidation of organics.

Other applications of DSC or PDSC oxidation tests included: radiation-damaged polyethylene, cellulose, thin film oxidation, medical poymers, and asphalts.

The attendees agreed that this symposium was rewarding and much knowledge was gained. They commented that the prsentations focused on areas that have not been previously discussed. Symposium cochairman, Alan Riga, suggested that another meeting be organized in two years at a future NATAS conference. The speakers and attendees agreed.

Rick Seyler, Kodak, summarized the symposium in the form of the following poem:

OIT BLUES
A SYMPOSIUM REVIEW
By R. J. Seyler

Don't you know its not thermodynamic?
Rather, OIT is really quite kinetic!

So when you do your OIT
It most likely won't be like me!

With so many ways to do the test
Precision within-lab will be the best.

When finally on a protocol we agreed
Eaxct control of temperature we'll need.

So now proceed as best we can
Just make sure its an aluminum pan.

Beware when other metal is present
Copper catalysis is a serious contaminant.

Other conditions like area, weight, and flow
Their influence we will need to know.

To switch the purge gas it was shown
One need not, for the BIAS is now known.

A last condition we yet must know
What point in time do we declare zero?

When all our tests we run one way
Which value of OIT do we report today?

Extrapolated onset, threshold, or the peak
'Tis a single number of value that we seek.

We need to realize that induction time
May have different significance for yours and mine.

Unless our sample is without formulation
OIT will address the type of stabilization!

We have tried our best to listen well
To learn all the messages you had to tell.

But when all is said and done
We should yet be troubled by just this one.

That from accelerated aging over which we toiled
The egg we hatched may have been Hard Boiled!

Alan T. Riga
The Lubrizol Corporation
    Wickliffe, OH; Symposium
    cochairman and coeditor

Gerald H. Patterson
The Lubrizol Corporation
    Wickliffe, OH; Symposium
    cochairman and coeditor

Oxidative Behavior of Polymers and Petrochemicals

Roger L. Blaine[1,2], C. Jay Lundgren[1] and Mary B. Harris[1]

OXIDATIVE INDUCTION TIME - A REVIEW OF DSC EXPERIMENTAL EFFECTS

REFERENCE: Blaine, R. L., Lundgren, C. J., and Harris, M. B., **"Oxidative Induction Time - A Review of DSC Experimental Effects,"** Oxidative Behavior of Materials by Thermal Analytical Techniques, ASTM STP 1326, A. T. Riga and G. H. Patterson, Eds., American Society for Testing and Materials, 1997.

ABSTRACT: Over the past several years, a number of ASTM committees have explored a wide variety of experimental parameters affecting the oxidative induction time (OIT) test method in an attempt to improve its intra- and inter-laboratory precision. These studies have identified test temperature precision as a key parameter affecting OIT precision. Other parameters of importance are oxygen flow rate, specimen size, specimen pan type, oxygen pressure and catalyst effects. The work of Kuck, Bowmer, Riga, Tikuisis and Thomas are reviewed as well as the collective work of ASTM Committees E37, D2, D9 and D35.

KEYWORDS: differential scanning calorimetry, oxidation, oxidative induction time, oxidative stability, polyethylene, polyolefins, thermal analysis

---

Oxidative Induction Time (OIT) is an accelerated aging test. It provides an index useful in comparing the relative resistance to oxidation of a variety of hydrocarbon materials. The test consists of heating a specimen to an elevated temperature (often 200 $^{\circ}$C) in a differential scanning calorimeter (DSC). Once temperature equilibrium is established, the specimen atmosphere is changed from inert nitrogen to oxidizing air or oxygen. The time from first oxygen exposure until the onset of oxidation is taken as the OIT value. This general procedure is applied, for example, to polyethylene wire insulation [1, 2], geosynthetic barriers [3], edible oils [4], lubricating oils and greases [5, 6]. Table 1 shows a few of the currently used application areas for OIT.

---

[1] Applications development manager, and applications chemists, respectively, TA Instruments, Inc., 109 Lukens Drive, New Castle DE 19720.

[2] Corresponding author.

Table 1 -- <u>Applications</u>

Material	Applications
Polyolefins	Wire and Cable Insulation
Polyethylenes	Pipe
Polyolefins	Geosynthetic Materials
Greases	Lubricants
Peanut Oil	Confections
Oils	Lubricants
Hydrocarbons	Fuels
Oils	Automatic Transmission Fluids

Most materials are tested to measure the effectiveness of the antioxidant package added to improve lifetime, although a few materials (e.g., edible oils) are tested in their natural, non-fortified state.

Background

The OIT procedure was first developed by Gilroy and coworkers at Bell Laboratory as a test procedure to screen polyethylene insulation used in telephone wire and cable for its oxidation resistance in pedestals [7]. The method first became available as a Western Electric specification [8] and later as ASTM Test Method for Copper Induced Oxidative Induction Time of Polyolefins [1]. Polyolefin manufacturers quickly embraced the procedure and began to apply it to other applications including raw resins, finished pipe [9], as well as to wire and cable insulation [2], and, most recently, geosynthetic waste pit liners [3].

It has long been known that the effectiveness of antioxidants, as measured by the OIT at high temperatures, may differ as a function of temperature [7, 10]. This may be due to a number of causes including changing mechanisms, loss of antioxidant due to volatilization at high test temperatures, etc. Many users would like to move the OIT test temperature closer to the actual use temperature to avoid some of these difficulties. Further, as additive packages have improved, OIT values have become progressively longer. In order to shorten the analysis time and to reduce test temperatures, increasing use is being made of Pressure DSC to accelerate the measurement at lower test temperatures.

The expanding applications for the OIT test method, the widespread use of Pressure DSC and improvements in apparatus electronics have combined to create increased interest in re-examination of the parameters of the OIT test method aimed at improving its precision. This interest has largely focus on improving inter- (between) laboratory reproducibility since intra- (within) laboratory repeatability is generally quite good. A general rule-of-thumb is that interlaboratory reproducibility should be about twice the within laboratory repeatability. For much OIT work, the reproducibility is much poorer than this rule-of-thumb. This indicates that within a single laboratory, the same thing is done the same way every time but that there are differences in procedure in going from one laboratory to another.

Several authors, and groups of workers have (re)examined the effects of a number of experimental parameters on OIT values and their precision. It is the purpose of this paper, then to review and report on the efforts of ourselves and others in there efforts to improve the OIT test method.

<u>Endpoint Selection</u>

The onset of oxidation is taken as the endpoint for the OIT measurement. Two means of determining the oxidation onset are in use. The most common is the "extrapolated onset" in which tangents are drawn at the point of maximum rate of oxidation and the baseline prior to the oxidation. Their intersection is taken as the endpoint for the OIT measurement.

The second method for establishment of the endpoint for the OIT determination is the point of "first-deviation-from-baseline". In this approach, some "threshold" is set above the baseline prior to oxidation (say 0.05 W/g). The endpoint for the OIT determination is taken at the point where the exothermic event crosses that threshold.

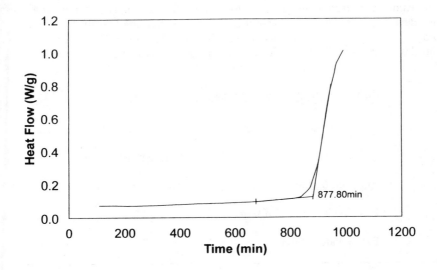

FIG. 1 -- Single stage oxidation endpoint determination.

If the oxidation exotherm is sharp, these two endpoint indicators produce similar results as seen in Figure 1 where the two values differ by only a few percent. Some materials, however, seem to have a multi-staged oxidation and the endpoint established by the two experimental procedures may be quite different as shown in Figure 2.

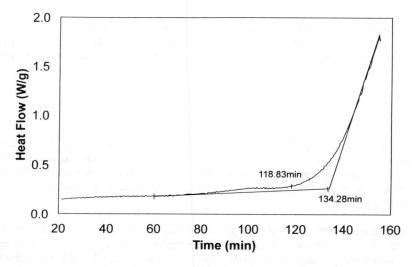

FIG. 2 -- Multistage oxidation endpoint determination

The selection of the method of determination of the OIT endpoint (i.e., extrapolated onset or first deviation) is the first parameter affecting the comparison of results from one laboratory to another. OIT values obtained by first deviation are usually lower than those obtained by extrapolated onset. Operators in a single laboratory commonly use the same

TABLE 2 -- Effect of endpoint selection on OIT precision.

Endpoint Method	OIT Standard Deviation (min)	
	Within Lab	Lab-to-Lab
First Deviation at 50 mW/g	4.5	7.3
Extrapolated Onset	2.8	7.2

approach, but workers in other laboratories may choose differing endpoint detectors. OIT results should identify the end point detector used to avoid this potential discrepancy.

Selection of the endpoint also affects the precision of the measurement. OIT precision, using extrapolated onset, usually has better precision than that using the first deviation from baseline. This is seen both in intralaboratory repeatability data and interlaboratory reproducibility values. ASTM's task group D9.18, working on an upgrade to ASTM D4565, obtained the OIT values for high density polyethylene insulation stripped from wire presented in Table 2. The OIT mean values for these tests were 122 and 126 min, respectively.

The within laboratory OIT precision values derived from the extrapolated onset is commonly two times better than that for the first deviation from baseline. For this reason, the extrapolated onset should be taken, wherever possible, as the endpoint indicator. The first deviation from baseline approach should be used only where this point on the oxidation profile provides specific information of interest to the researcher.

Effect of Temperature

The single most important influence in comparing OIT values from one laboratory to another is the test temperature of the method. Table 3 shows the effect slight changes in temperature can have on the OIT value for a high density polyethylene sample.

TABLE 3 -- Effect of temperature.

Temperature ($^{\circ}$C)	OIT (min)
198.0	40.8
200.0	35.7
202.0	29.2

For this polyethylene sample, the effect of temperature on the OIT value is 2.9 min / $^{\circ}$C or about 8.1 % / $^{\circ}$C at the 200 °C test temperature. If the test temperature is not exactly the same in two laboratories or is not that called for in the test method, a serious discrepancy is likely to result in comparing the results. This strong effect of test temperature is well known by those designing a test method [7] but may be ignored by the technologist who run the experiments because it requires instrument recalibration under conditions (i.e., isothermal operation) different than those used for most DSC experiments (e.g., 10 °C/min heating rate).

Precise temperature calibration along with the direct measurement and recording of sample temperature during the test are keys to overcoming lab-to-lab variability in OIT measurements. Some thermal analyzers have a strong temperature dependence on heating rate [11]. For this reason, temperature calibration must be carried out using a very low heating rate, usually 1 °C/min, to better simulate isothermal conditions.

Effect of Oxygen Flow Rate

A third experimental parameter which can effect the OIT value and its precision is the availability of the oxygen reactant. One of the factors affecting oxygen availability is flow rate.

OIT values are not strongly dependent upon reactant gas flow rates provided a necessary minimum flow rate is available. Unfortunately, this minimum level is quite close to the 50 mL/min commonly used [12]. Table 4 provides information on the effect of oxygen flow rate on the absolute value for OIT and on its precision, respectively, for a high density polyethylene sample.

TABLE 4 --Effect of oxygen flow rate on OIT.

Flow Rate (mL/min)	OIT (min)	Precision (%)
20	64	---
50	45	4.5
100	34	2.3

These results, and those of Ashby [12], indicate that around 50 mL/min flow rate, the OIT value may change 3.5 minutes (ca. 8%) for each 10 mL/min change in purge gas flow rate.

Further information presented in Table 4 shows that precision improves with higher flow rates. Unfortunately, not all DSC instruments are capable of flow rates as high as 100 mL/min so most standard methods have settled on the 50 mL/min rate for instrument compatibility purposes. The flow rate must be limited, however, to a very narrow range of ± 5 mL/min, to improve both repeatability and reproducibility.

Floating ball type flowmeters are commonly used in the thermal analysis laboratory for the measurement and control of purge gas flow rates. Flowmeter readings, however, are dependent on the molecular weight of the gas measured and must be calibrated for each gas used. Calibration tables for common gases are supplied with most high quality flowmeters and the use of such calibration tables with oxygen is essential. Flowmeter inadvertently set on the commonly used nitrogen gas setting will produce a lower than expected flow rate with oxygen leading to longer than anticipated OIT values. For example, a setting which reads 50 mL/min for nitrogen produces a 43.1 mL/min flow of oxygen. This is outside of the permitted range and is sufficient to change the indicated OIT value by approximately 5%.

Interlaboratory imprecision due to reactive gas availability effects can come from a second source. This is the difference among instrument designs in directing the reactant gas onto the test specimen. In some instruments, the reactive purge gas blanketing of the test sample is ineffective in reaching the test specimen [13]. Users should contact their instrument suppliers for any purge gas modifications required to achieve high OIT precision. Specifically ASTM D4565 states, "in some power compensation DSC's, use of the two-hole platinum sample holder lids with a special "flow through" swing away block cover is recommended" in order to obtain accuracy's and precision equivalent to other unmodified instrumentation" [9]. Users should follow all such instructions in order to obtain the best reproducibility.

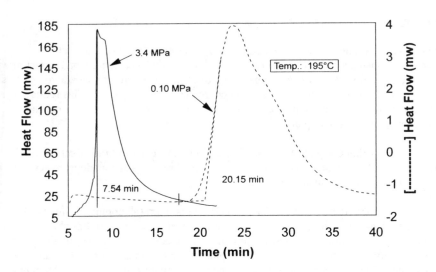

Fig. 3 -- Effect of oxygen pressure on OIT [14]

Effect of Oxygen Pressure

As in any oxidation reaction, the concentration of reactant oxygen has an effect on the OIT. Different OIT values will be obtained if air (with its 21% oxygen concentration) rather than pure oxygen is used as the reactive gas. Today, virtually all of the OIT test methods specify oxygen as the reactant gas.

The oxidation reaction rate can further be increased (and OIT values shortened) by using a Pressure DSC and elevated pressures of oxygen. Figure 3 illustrates, the reduction in OIT value resultant in operating at elevated oxygen pressure. In this case, the OIT of a lubricating grease is 20 minutes at atmospheric oxygen pressure, but is reduced to 7.5 minutes at 3.4 MPa (500 psi) pressure [14].

A number of workers, including Dugan [15], Tikuisis [16], and Thomas [17] have explored the effect of pressure on OIT values and have observed an Arrhenius type expression for partial pressures in excess of about 1.4 MPa (200 psi) with best results being obtained between 3.4 and 5.5 MPa (500 to 800 psi). The higher pressures are advantageous due to their insensitivity to small pressure changes and to the reduction in analysis time. In addition, higher pressures permit (lower) test temperatures to be used which are closer to actual end use conditions. For these reasons, virtually all of the pressure OIT work is done at pressure of 3.4 MPa or higher.

Effect of Catalysts (Sample Pan Materials)

The oxidation reaction is very sensitive to catalytic effects. It is well known that polyethylene insulation on copper wire has much lower OIT values than on aluminum wire or for the insulation alone. The original methods took advantage of the catalytic effects of copper oxide by running the test method in copper sample pans [1, 8]. This reduces the overall test time but induces interlaboratory imprecision due to different sources and conditioning of the copper pans used.

At least two workers, Riga [14, 18] and Kuck [13], have explored the effects of metal catalyst on the OIT value. Their work, based on DSC sample pans made from several different materials, is summarized in Table 5.

These results show that copper, stainless steel and platinum (all commercially available sample pans), catalyze the oxidation reaction of the test specimen. Standard commercial aluminum and non-commercial nickel pans show no catalytic effects. To obtain reproducibility, catalyst effects should be avoided through the use of standard aluminum sample pans. The price paid for this reproducibility is longer analysis times. Aluminum pans are required by several ASTM procedures [2,9].

Some operators make use of a stainless steel screen to keep the test specimen in place in the DSC pan by crimping it in place as a pseudo-lid. (Note that OIT measurements are to be run in open pans with no lid in place.) Stainless steel, however, serves as a catalyst in the oxidation reaction. Workers who wish to make use of the screen principle should use an aluminum screen rather than the stainless steel. Use of stainless steel is likely to give shorter OIT values with poorer precision .

TABLE 5 -- Metal catalysis effect on polyethylene OIT.

	OIT (min)	
Metal	Case 1[14,18]	Case 2 [13]
Aluminum	47.9	36.5
Copper	21.4	----
Stainless Steel	38.9	26.9
Nickel	----	36.6
Platinum	----	30.5

Effect of Sample Mass and Form

Since OIT is a measure of a chemical reaction in which one of the reactants is a gas surrounding a condensed phase test specimen, the surface area of the specimen exposed to the oxygen is likely to be an important experimental parameter. The most precise OIT methods call for small (usually < 5 mg) and uniform (± 0.1 mg) test specimen sizes. If larger specimens are used, an appreciable portion of the specimen is shielded from the

reactive gas. Ashby showed that test specimens larger than about 5 mg had a relatively constant OIT value and argued for a selection of sample size in this region to minimize sample mass effects [12]. He further showed that specimens with exposed areas greater than 50 mm^2 appeared to have little change in OIT value with increased surface area. These results argue that there is, indeed, a surface area effect which must be considered.

Rhee illustrated for grease samples that specimen of the same weight but differing surface area produced markedly different OIT values as shown in Figure 4 [5]. He recommends that for liquid specimens, that Solid Fat Index (SFI) sample pans be used to produce a uniform sample area exposed to the oxygen when small specimen weights are used [6].

Specifying a large sample (to take advantage of mass and area insensitivities) is complicated by the fact that specimen size and shape often change upon heating through the melt. Polyolefin films, formed to obtain the maximum surface area for the minimum mass, tend to "bead up" into a ball upon melting. In this case, a uniform specimen size is more important than the test specimen weight.

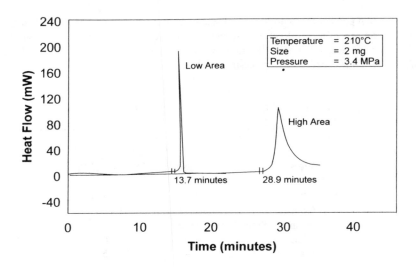

FIG. 4 -- Effect of area on OIT [4]

Effect of Time

The repeatability standard deviation for the OIT measurement is nearly constant in the 2 to 3 minute range and is not strongly dependent upon the OIT value itself. This means that precision should be stated in the form of standard deviation and not as percent relative standard deviation. This has the effect of dramatically increasing the relative standard deviation as the OIT values decrease. For example, at 3 min standard deviation becomes a 20% relative standard deviation for a 15 minute OIT value but is only 4% for a 75 minute OIT value

Because of this effect, experimental conditions should be selected to that the OIT values is longer than 15 minutes. Values larger than 15 minute will permit reasonable relative standard deviation values to be obtained. At the upper end, experimental conditions are often selected so that the maximum OIT values are less that 100 minutes to improve productivity.

While the repeatability and reproducibility of the OIT test method will have some dependence on the sample itself, the best available within laboratory repeatability standard deviations (68% confidence limit) is about 1.6 min. and lab-to-lab reproducibility is about 3.1 min. Most ASTM test methods use the 95% confidence limit, 2.8 times the standard deviation. OIT results, each the mean of duplicate determinations, should be considered suspect if the within laboratory repeatability is greater than 4.5 min and interlaboratory reproducibility is greater than 8.7 min (in the best case).

Summary

The OIT method is very sensitive to a number of experimental conditions including temperature, oxygen flow rate, method of onset determination, specimen mass and surface area. To obtain the best precision, experimental conditions must be maintained within a very narrow set of conditions. Under such carefully controlled conditions, repeatability as good as 4.5 min and reproducibilities as good as 8.7 minutes may be obtained.

The most important effect on reproducibility is calibration of the apparatus for the identified test temperature. This parameter has a bias of 8 % / $^{\circ}$C in OIT value. The second most influential parameter on both interlaboratory reproducibility and intralaboratory repeatability is the flow rate of the reactive oxygen used in the test method. A bias on the order of 8 % / 10 mL/min change in flow rate may result. Calibration of the flow measuring apparatus for oxygen use is essential.

Acknowledgments

Thanks are given to the many workers, cited and non-cited, who contributed to this effort. In some cases, the contributions of individual scientists and organizations were cited here, but much of the reported work took place in task groups. Many individuals from a large number of organization contributed in the this way. A few of the organization which contributed to this work are listed Table 6.

TABLE 6 -- Organizations studying oxidative induction time.

o ASTM Committee D2 on Petroleum Products

o ASTM Committee D9 on Wire and Cable Insulation

o ASTM Committee D20 on Plastics

o ASTM Committee E37 on Thermal methods

o ASTM Committee D35 on Geosynthetics

o Institute of (British) Petroleum

o National Institute for Standards and Technology

REFERENCES

[1]   ASTM D3895, "Test Method for Oxidative Induction Time of Polyolefins by DSC", American Society for Testing and Materials.

[2]   ASTM D4565, "Test Method for Properties of Insulations and Jackets for Telecommunications Wire and Cable", American Society for Testing and Materials.

[3]   ASTM D5885, "Test Method for Oxidative Induction Time of Polyolefin Geosynthetics by High Pressure DSC", American Society for Testing and Materials.

[4]   Kowalski, B., Thermochim. Acta, 156, 347-358 (1989).

[5]   Rhee, In-Sik, NLGI Spokesman, 55, 7-123 to 16-132 (1991).

[6]   ASTM D5483, "Test Method for Oxidation Induction Time of Lubricating Greases by Pressure DSC", American Society for Testing and Materials.

[7]   Howard, J.B., and Gilroy, H.M., Polym. Eng. Sci., 15, 268-271 (1975).

[8]   Western Electric Manufacturing Standard 17000, Section 1230 (1971)

[9]   ASTM D3350 "Specification for Polyethylene Plastic Pipe and Fitting Materials", American Society for Testing and Materials.

[10]   Hsuan, Y.G., "Evaluation of Oxidation Behavior of Polyolefin Geosynthetics Using PDSC", Oxidative Behavior of Materials by Thermal Analytical Techniques, ASTM STP 1326, A.T. Riga and G.H. Patterson, Eds., American Society for Testing and Materials, 1997.

[11]   ASTM E0698, "Test Method for Arrhenius Kinetic Constants for Thermally Unstable Materials", American Society for Testing and Materials.

[12]   Ashby, R.T., Proc. 3rd Intern. Conf. Plastics in Telecommunications, 25-1 to 25-11.

[13]   Kuck, V.J., Proc. 6th Intern. Conf. Plastics in Telecommunications, 3-1 to 3-10 (1992)

[14]   Patterson, Gerald H., and Riga, Alan T., Thermochim. Acta, 226, 201-210 (1993).

[15]   Dugan, George, and McCarty, J.D., Proc. 13th North Amer. Therm. Anal. Soc. Conf., 520-525 (1984).

[16]   Tikuisis, T., Lam, P., and Cossar, M., Proc. 6th Seminar on Geosynthetics, Geosynthetic Research Institute, Drexel University, Philadelphia PA, 191-201 (1992).

[17]   Thomas, R.W., and Ancelet, C.R., Proc. Geosynthetic Conf., 2, 915-924 (1993).

[18]   Riga, A.T., Proc. 20th North Amer. Therm. Anal. Soc. Conf., 517-521 (1991).

Joseph M. Perez[1]

OXIDATION STUDIES OF LUBRICANTS FROM DORNTE TO
MICROREACTORS

REFERENCE : Perez, J. M., " Oxidation Studies of Lubricants from Dornte to
Microreactors ," Oxidative Behavior of Materials by Thermal Analytical Techniques,
ASTM STP 1326, A. T. Riga and G. H. Patterson, Eds., American Society for Testing
and Materials, 1997.

ABSTRACT : Friction, wear and the oxidation of lubricants are concerns in most
mechanical systems.   Selection of the best current thermoanalytical method to evaluate the
oxidative performance of a lubricant for an application is an ongoing problem. The
variability and complexity of mechanical systems makes correlation with bench tests
difficult. The oxidative environment in automotive applications may vary from nil in
sealed-for-life components to severe oxidative conditions found in piston, ring, and liner
zones. The researcher's task is to develop useful and meaningful bench tests to allow the
engineer to concentrate his effort and resources on developing better lubricants for a given
application.   This paper reviews the evolution of oxidation test methods for the past 50
years from large complex systems to the current microanalysis methods. The mechanism
of oxidation has changed very little but our understanding of additive interactions has
advanced significantly.   Additive depletion and the resulting oxidation process can be
studied using the microthermooxidative methods. Often, a combination of methods can be
used to obtain a better understanding of the thermal and oxidative processes occurring.
Some comparisons of the methods and their applications are described.

KEYWORDS: oxidation, lubricants, thermaloxidative methods, thermoanalytical
techniques, microoxidation

[1]Adjunct Professor, Chemical Engineering Department, The Pennsylvania State
University, University Park, PA 16802.

The evolution of thermoanalysis methods over the past 50 years has resulted in significant improvements in how we evaluate lubricant oxidative and thermal stabilities. To contrast today's methods with those used 50 years ago, this paper compares some of the techniques used in the 1940's and 1950's to evaluate oxidation of lubricants and some of the newer techniques. These include the differential scanning calorimetry (DSC), thermo-gravimetric analysis (TGA), Fourier transform infrared analysis (FTIR), the Klaus Penn State microreactor (PSMO), and related supportive methods. All of these tests are rapid and require only microquantities of material for analysis. However, with the use of these microanalysis techniques (MATs), comes the risk of misinterpretation of the results relative to fullscale tests, field applications, and performance. Computer and electronics technologies have helped minimize instrumental systems and, more significantly, collect and interpret the data from the systems. To take advantage of the new technologies only requires the appropriate skills. The key to the successful use of newer analytical tools is in the decision-making process, with the final decision left to the analyst. This decision may be as simple as to whether to accept or reject computer results. To do this correctly requires an understanding of the complexities of the analytical system used and a knowledge of the related thermal and oxidative processes.

HISTORICAL OVERVIEW

Mechanism of Oxidation

The theories of liquid phase oxidation in which air or oxygen is the oxidizing agent go back over a hundred years. Bone in 1898 [1] developed one of the first theories in which hydroxylation and thermal decomposition were claimed to be the key drivers in the process of oxidation. Oxidation occured by the formation of unstable hydroxylated molecules that thermally decomposed into simpler, more stable oxygenated products (Fig. 1).

1) Formation of unstable hydroxylated molecules:

$$CH_3\text{-}CH_3 \longrightarrow CH_3\text{-}CH_2\text{-}OH \longrightarrow CH_3\text{-}CH(OH)_2$$

2) Thermal decomposition:

$$CH_3\text{-}CH(OH)_2 \longrightarrow CH_3\text{-}CHO + H_2O$$

3) Formation of other products:

$$CH_3\text{-}CHO \longrightarrow CH_3\text{-}C(O)\text{-}OH + CH_2(OH)C(O)\text{-}OH + CH_3OH + CO_2$$

FIG.1 -- Bone oxidation theory, circa 1898.

Although his theory was accepted by some researchers for a number of years, Bone's theory failed because the mechanism did not account for the formation of peroxides. It was unlikely that the primary intermediate products were alcohols due to their slow rate of oxidation. The peroxides were identified by others as one of the predominant early products. Lewis [2] in 1926 suggested that the initial step was dehydrogenation in which oxygen reacted with the most loosely bound hydrogen producing water and unsaturated hydrocarbons which later reacted to form peroxides, aldehydes, and other products. The theory was based on vaporphase oxidation and was suggested to occur in the liquid phase also. Lewis' theory was not supported by experimental evidence and received little acceptance.

Bach [3], Engler and Wild [4] suggested a peroxide theory in 1887 in which the initial step in lowtemperature oxidation is the formation of peroxides. This theory was further developed by Callander while studying the oxidation of paraffins in 1927 [5] and Ubbelohde in 1935 [6]. Zuidema [7] in 1946 reviewed the work of various investigators and proposed a similar mechanism. His theory included the initial formation of a hydroperoxide followed by decomposition to a ketone or an alcohol, followed by further oxidation to form aldehydes, acids, oxides of carbon, and water. George et al [8 , 9] proposed a similar mechanism but suggested ketones were the predominant intermediate. Numerous authors have studied and refined the oxidation theories but by the 1950's, the oxidation mechanism was well established (Fig. 2).

Start:
$$RH \longrightarrow R\bullet \tag{1}$$

Chain popagation:
$$R\bullet + O_2 \longrightarrow ROO\bullet \tag{2}$$
$$ROO\bullet + RH \longrightarrow ROOH + R\bullet \tag{3}$$

Chain branching:
$$ROOH \longrightarrow RO\bullet + \bullet OH \tag{4}$$
$$RO\bullet + RH \longrightarrow ROH + R\bullet \tag{5}$$
$$\bullet OH + RH \longrightarrow H_2O + R\bullet \tag{6}$$

Chain termination:
$$2R\bullet \longrightarrow R\text{-}R \tag{7}$$
$$R\bullet + ROO\bullet \longrightarrow ROOR \tag{8}$$
$$2 ROO\bullet \longrightarrow ROOR + O_2 \tag{9}$$

FIG. 2 -- Current oxidation theory.

Two things are significant in all of the work before 1950: first, instrumental methods were very limited, even crude at the best, compared to today's standards, and second, it took about 20 to 25 years for each step forward in the evolution of thermoxidative processes. This evolution continues. In the twenty years after the 1950's, oxidation studies were dominated by new and improved instrumental methods such as gas chromatography(GC) and mass spectroscopy(MS). The most recent 20 years, have through necessity, environmental concerns, and increasing costs of doing research, resulted in the development of useful micromethods of analysis.

Oxidation Systems and Analysis Methods

The evolution of methods used to study oxidation in the laboratory are reviewed, (Fig. 3). Early methods included both closed and open systems. The closed systems were essentially sealed bombtype tests in which the lubricant was sealed in some type of steel bomb, heated in a controlled manner, cooled, and the contents analyzed. These methods are the predecessors of more refined current methods such as the ASTM Test Method D#2272 for the Determination of Oxidation Stability of Steam Turbine Oils by Rotating Bomb (RBOT) [10] and ASTM Test Method D#4742 for Oxidation Stability of Gasoline Automotive Engine Oils by ThinFilm Oxygen Uptake test. In the open systems, air or oxygen were passed over the surface or bubbled through the lubricant under controlled conditions. The goal in these systems was to measure the quantity of oxygen used and to somehow trap and analyze the products of oxidation.

LUBRICANT TEST METHOD EVOLUTION:

1900	**CONDENSORS + WET CHEMISTRY**
1936	**DORNTE APPARATUS**
1940	**DORNTE UNITS**
1958	**MODIFIED DORNTE (PRL)**
1960	**BULK CLOSED AND OPEN SYSTEMS**
1978	**PENN STATE MICROREACTOR**
1980	**DIFFERENTIAL SCANNING CALORIMETER**
1983	**THERMOGRAVIMETRIC ANALYSES**
1984	**THIN FILM OXYGEN UPTAKE TEST**
1990	**PRESSURIZED DSC**

FIG. 3 -- Evolution of oxidation tests

In the early 1900s, most research involved the condensation of vapor phase oxidation products followed by wet chemistry analysis to obtain data on the processes. This was used to support the oxidation theories of the time. The systems improved and the most widely used apparatus in the 1940s was first employed by Dornte in 1936 [11]. A modification of this apparatus was used by Booser [12], Shirk [13] and others at the Petroleum Refining Laboratory [PRL] at The Pennsylvania State University. The apparatus shown in Fig. 4 was described by Fenske et al. in 1941 [14].

The two sections displayed in the figure were actually connected and sat side by side. The left hand (upper) section comprises the oxygen supply, an "automatic" oxygen makeup device, and a means to monitor and record the volume of the gas flow. The right hand (lower) section includes an oxygen reservoir, a circulating pump, an all glass oxidation reaction tube, and a system of traps and adsorbers to remove volatile oxidation products from the gas stream before returning the oxygen stream to the reservoir. The manpower requirements to keep the systems operating were excessive. Maintenance of plugged traps, leaks, and other trouble spots kept the operator busy round the clock. In addition, considerable effort was required to analyze the products recovered from the system after the test was completed.

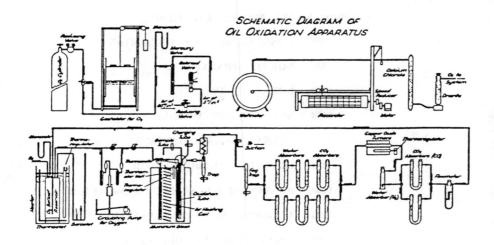

FIG. 4 -- Dornte oxidation apparatus.

A simplified version of the system was devised in the late 1950s [15] and used to study differences in the oxidation of conventional and superrefined mineral oils [16]. The size of the system (Fig. 5) was reduced, but most of the analyses were still manpower intensive wet chemistry methods. Gas chromatography was a novel tool and just starting to be used for routine analysis at the completion of this work. The Orsat method [17] was used to analyze the gaseous and lower molecular weight products. The study resulted in simplified bulk oxidation and thermal tests, in which, oxygen consumption and trapped reaction products could be analyzed.   It became obvious as the result of oxidation studies over several years that the main limitation of this type of system, or similar systems used at the PRL, was that the oxidation rate was controlled by oxygen diffusion. A number of techniques including smaller volume tests, tests with increased catalyst surface area (coiled wire, stacked metal specimens, and various lengths of chain), panel cokers, concentric tubes with an annulus for lubricant and oxygen flow, and systems with controlled flow through heated tubes were used to improve the ability to predict the expected performance in engines and turbines. The not so obvious solution was the use of thin film methods.

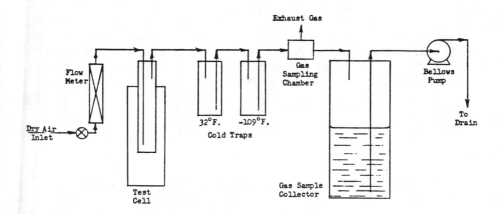

FIG. 5 -- Modified Dornte apparatus.

THIN FILM OXIDATION METHODS

Thin film methods can be used to produce data on the oxidation of lubricants in a few hours, using milligram or microlitre quantities of the lubricant, compared to weeks of testing in an apparatus that required a couple of hundred millilitres or as much as several gallons of the lubricant to perform a test series.

Three thin film methods that evolved from earlier studies have been used extensively with varying degrees of success. These include the Thin Film Oxygen Uptake Test (TFOUT) method developed by NIST [18,19], the Klaus Penn State Microreactor (PSMO) [20,21] and the Differential Scanning Calorimeter (DSC) [22,23].

TFOUT

The TFOUT is a well established ASTM Test Method D#4742 and was developed to correlate with the IIID Engine Oil Test Method [18]. More recently, the test was modified to better meet the newer IIIE test modification [19]. Basically, under an oxygen atmosphere and controlled conditions, the TFOUT measures additive depletion and the onset of oxidation as indicated by oxygen uptake as measured by pressure drop.

PSMO

The PSMO reactor has been used to predict lubricant performance of IIID oils [24] and diesel engine deposit forming tendencies of lubricants [25,26]. The test has been used to study oxidation mechanisms of mineral oils [27], synthetic lubricants [28,29] and more recently to study the oxidation of vegetable oils and greases.   The basic method has been thoroughly described in the references and involves the oxidation of a 20 or 40ul thin film of oil, or grease, under controlled conditions. The test yields information on evaporation and deposits by mass. Gel permeation chromatography (GPC) is used to follow the oxidation mechanisms by determining the relative amount of material of different molecular size.   The test time can vary from a few minutes to several hours depending on the test temperature and stability of the sample.

DSC

The DSC is routinely used to study additives and base fluids by a number of laboratories. The high pressure version of the test, the Pressurized DSC (PDSC), was used to develop lubricants for advanced diesel engines [30].   The method is used to evaluate additive depletion and rates of oxidation by observing the exothermic and endothermic reactions occurring when a thin film of oil is subjected to a pressurized air or oxygen atmosphere at different temperatures.

The "twopeak method" is a programmed temperature version of the PDSC test and was used to evaluate deposit forming tendencies at high temperatures [31,32].   The "twopeak method" was developed to study top land carbon forming tendencies in diesel engines, and the PSMO tends to simulate crankcase atmospheres and longterm oxidation effects at intermediate temperatures [33].

FTIR

The FTIR can be used to complement the thinfilm analysis methods.   The micro-FTIR instrument has been used to follow oxidation product formation in the thin film tests. Analysis of selected bands can be used to determine the increase in oxidation with time in the systems. Combinations of methods in oxidation studies can result in significant amounts of data in short periods of time.

TGA

The TGA method uses microquantities of liquid to obtain both thermal and oxidative data on tests fluids. The method is useful in obtaining volatility data under various atmospheres. It is another method that is complementary to the oxidation tests .

EXPERIMENTAL METHODS and TEST RESULTS

Several of the experimental methods to study oxidation are compared (Table 1). The methods evolved along with a better understanding the mechanism of oxidation. The advantages of the newer methods are less time required to conduct tests, higher application temperatures and better analysis instrumentation. Some laboratory test results obtained with the various methods are described.

TABLE 1 -- Comparison of methods.

Apparatus	Sample Size	Environment	Temp., °C	Test Time	Anal. Method / Comments
Dornte	300-500 g	Oxygen	<180	Month	Wet Chemical
Mod. Dornte	100-200 g	Air/Nitrogen	175 to 260	Weeks	" " + Orsat
TFOUT	1.5 g	O₂ (696 kPa)	160	Hours	None/Catalysts
TGA	10 mg	Air/ O₂ /Inert	*	Minutes	Volatility
PDSC	1 to 3 mg	Air/ O₂ /Inert	*	Minutes	FTIR/Additive
PSMO	20 to 40 ʋg	Air/ O₂ /Inert	175 to 275	1 to 4 Hours	GPC/FTIR

Bulk Oxidation Tests

The oxidation test method used depends on the information required. The bulk oxidation tests do produce ample quantities of fluid for further analysis and correlations with applications in the field may require such testing. The study comparing the oxidation of conventional and superrefined mineral oils [15] which included liquid column chromatographic separation of the products, required months of testing and over a gallon of each fluid tested.   The tests demonstrated differences in the oxidation of conventional lubricants containing natural inhibitors and highly refined oils from which the natural inhibitors were removed. The PSMO test can supply most of the data in several days. However, it would require a combination of analytical methods to conduct the analyses.

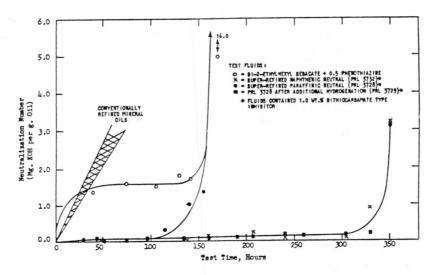

FIG.6 -- Bulk oxidation studies of additives.

Although cumbersome and time-consuming, running the early studies did provide significant information on additive depletion (Fig.6). The effectiveness of additives in different base stocks depended on the degree of refining or the type of synthetic ester used. The oxidation process was evaluated using a combination of wet methods, column chromatography, and infrared analyses. In general, the oxidation attack on the conventionally refined mineral oils was concentrated on natural inhibitors. Table 2 contains some of the analysis results. The percentage of the original molecules that are oxidized for a given degree of oxidation increased with increased or superrefining and increasing temperature. The number of oxygenated linkages per molecule of oxidized oil decreased in the more highly refined stocks. Volatile products, including water, accounted for as much as 68 to 86 % of the assimilated oxygen.

TABLE 2 -- Oxidation of paraffinic mineral oils.

Oxidation Temp, °C	Moles O_2 Abs. per Mole Oil	Fluid Loss, wt %	Unoxidized Fluid, wt %	Wt % Fluid Affected by Oxidation	Wt % Oxid. Liquid Product	Avg. Distribution of Oxy-groups Per Molecule
-------------- Conventionally Refined Oil --------------						
260	0.92	7	62.2	37.8	30.8	1.4
175	0.92	5.6	68.8	31.2	25.6	1.5
-------------- Hydrogenated (Super-Refined) --------------						
260	1.08	15.7	53.1	46.9	31.2	1.1
175	1.21	9.1	57.6	42.2	33.3	1.6

PSMO Tests

Volatility -- The PSMO has been used to evaluate mineral oils, synthetic lubricants and natural lubricants. The method is used to determine the relative volatility of the lubricants under an inert or an oxidizing atmosphere. The relative volatility of several lubricants are shown, Table 3. In this case, the relative volatility was conducted under an oxidizing atmosphere. The losses are determined by weighing the test specimens before and after the tests. The ability of additives to reduce oxidative volatility can also be studied. The effect of an oxidation inhibitor package on a synthetic oil is shown. To determine the relative rates in an inert atmosphere simply requires the use of argon or nitrogen in place of the air in the tests. The effect of the substrate can also be studied.

Oxidation -- The relative oxidation stability can be determined in the PSMO by dissolving the remaining oxidized liquid in a solvent and conducting an appropriate analysis. In most studies conducted, the GPC has been used to determine the distribution of oxidation products. The high molecular weight fractions are the precursors to the formation of deposits in the test. To determine the unreacted fraction of the original material [34] requires a GPC analysis of the solution of oxidized fluid before and after passing it through a bauxite or silica gel cartridge to remove the oxidized material in the same molecular weight range as the original fluid. An example of GPC evaluations of an oxidized vegetable oil are shown in Fig. 7.
The amount of deposits formed by oxidation of the oil are determined by weighing the residue on the test specimen after removing the liquid products. The deposits can range from less than 0.1 to over 60 weight percent of the initial fluid charged depending on test conditions and the fluid evaluated. The effect of metals on the rate of oxidation has also been studied by use of different materials as the test specimen[35].

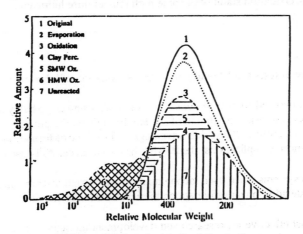

FIG.7 -- GPC analysis of oxidized oil.

PDSC Tests.

The PDSC test can be used to evaluate the effectivenes of antioxidant additives (Fig. 8), or the rate of oxidation of lubricants after the depletion of additives. An evaluation of the oxidized product can be made by terminating the test at any point and analyzing the residue using Fourier Transform Infrared (FTIR). The sample remaining after an isothermal evaluation can be evaluated using techniques similar to those used for the PSMO. The mechanisms in the PSMO and PDSC tests can be significantly different, especially if the PDSC is operated in a programmed temperature mode.

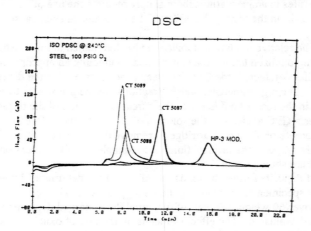

FIG. 8 -- Oxidation stability of some high temperature lubricants.

Summary

- There are a variety of tests available to study thermooxidation of lubricants.

- The methods have evolved from time consuming methods using large bulky systems to rapid microanalysis systems. The choice of system depends on the purpose of the testing. To screen additives, PDSC has advantages. For deposit formation in diesel and gasoline automotive applications, the PSMO has been successfully used.

- The microanalysis systems when used inconjunction with current instrumental analysis systems can provide additional useful data on lubricant oxidative stability..

- Bench tests are cost effective for research and development lubricants. Screening of lubricants prior to testing in more costly engine or field tests does not replace fullscale engine or field tests. However, bench tests can be effectively used to reduce the number of oils evaluated in the expensive, time consuming tests.

Acknowledgments

This work is dedicated to past members of the Petroleum Refining Laboratory at Penn State University for their pioneering efforts on oxidation and to the more recent members of the Tribology Group under Professor Emeritus E. E. Klaus for their continuing efforts to understand the oxidation mechanisms of lubricants. Current members contributing data to this paper include R. Adams, P. Gabliondo, D. Weller and S. Asadauskas.

References

[1] Bone, W. A. and Drugman, J., J. Chem. Soc., Vol. 89, 1906, p. 660.

[2] Lewis, J. L., J.Chem.Soc. Vol.110, 1927, p.1555.

[3] Bach, A., Compt. Rend., Vol.124, 1987, p. 951.

[4] Engler, C. and Wild, W., Ber., Vol.30, 1897, p.1669.

[5] Callender, H. L., Engineering, Feb.1927.

[6] Ubbelohde, A. R., Proc.Roy. Soc., Vol. A152, 1935, p.354.

[7] Zuidema, H. H., Chem. Rev., Vol. 38, 1946, p.197.

[8] George, P., Rideal, E. K. and Robertson, A., Nature, Vol.149, 1942, p.601.

[9] George, P. and Walsh, A. D., Trans.Far. Soc.,Vol. 42, 1946, p.94.

[10] Von Fuchs, G. H., Claridge, E. L. and Zuidema, H. H., "The Rotary Bomb Oxidation Test for Inhibited Turbine Oils," MTRSA (formerly ASTM Bulletin), No. 186, December (1952).

[11] Dornte, R.W., Ind. Engrg. Chem., Vol.28, No.26, (1936).

[12] Booser, E. R., Ph.D. thesis, The Pennsylvania State University, University Park, PA, 1948.

[13] Shirk, N. E., Ph.D. thesis, The Pennsylvania State University, University Park, PA, 1953.

[14] Fenske, M. R., Ind. Engrg. Chem., Anal.Ed., Vol.13, No.51, (1941).

[15] Perez, J. M., M.S. thesis, High Temperature Oxidation of Mineral Oils, The Pennsylvania State University, University Park, PA 1959.

[16] Perez, J. M., Klaus, E. E., and Fenske, M. R., "Development and Use of a Simplified Quantitative Oxidation Test for Lubricants," ACS, Div of Pet. Chem, Cleveland, OH, April 1960, .

[17] Matuszak, M. P., The Gas Analysis Manual, Fisher Scientific Co., Pittsburgh, PA, c.1950

[18] Ku, C. S. and Hsu, S. M., "A Thin Film Oxygen Uptake Test for the Evaluation of Automotive Lubricants," Lubr.Engrg., Vol. 40, No. 2, 1984, pp.75-83.

[19] Ku, C. S., Pei, P. and Hsu, S. M., "A Modified ThinFilm Oxygen Uptake Test (TFOUT) for the Evaluation of Lubricant Stability in ASTM Sequence IIIE," SAE Technical Paper 902121, Tulsa, OK (1990).

[20] Citkovic, E., Klaus, E. E. and Lockwood, F., "A Thin-Film Test for Measurement of the Oxidation and Evaporation of Ester-Type Lubricants," ASLE Trans., 22, 1979, p.395.

[21] Klaus, E. E., Krisnamachar, V.and Dang, H., "Evaluation of Basestock and Formulated Lubes Using the Penn State Microoxidation Test," NBS Spec. Publ. No. 584, 1980,p.285-294.

[22] Walker, J. A. and Tsang, W., Characterization of Oils by Differential Scanning Calorimetry, SAE Technical Paper 801383 (1980)

[23] Hsu, S. M., Cummings, A. L., and Clark, D. B., "Evaluation of Crankcase Lubricants by Differential Scanning Calorimetry," SAE Technical Paper 821252, (1982).

[24] Klaus, E. E., Cho, L. and Dang, H., "Adaption of the Penn State Microoxidation Test for the Evaluation of Automotive Lubricants," SAE Paper 801362, SAE SP.Publ. 80473, p.83-92, San Francisco,CA, (1980)

[25] Perez, J. M., Kelley, F. A., Klaus, E. E., and Bagrodia, V., "Development and Use of the PSU Microoxidation Test for Diesel Oils," SAE Technical Paper No. 872028 (1987).

[26] Zerla, F. N. and Moore, R. A., "Evaluation of Diesel Engine Lubricants by Microoxidation," SAE Technical Paper 890239 (1989).

[27] Wang, C. C., Duda, J. L. and Klaus, E. E., "A Kinetic Model of Lubricant Deposit Formation Under ThinFilm Conditions," Tribology Trans., 37, 1994, p.168-174.

[28] Ali, A., Lockwood, F., Klaus, E. E., Duda, J. L. and Tewksbury, E. J., "The Chemical Degradation of Ester Lubricants," ASLE Trans., 22, 1979, p. 267-276.

[29] Cho, L. and Klaus, E. E., "Oxidative Degradation of Phosphate Esters," ASLE Trans., 24(2),1981, p.276-284.

[30] Perez, J. M., Ku, C. S. and Hsu, S. M., "High Temperature Liquid Lubricant for Advanced Engines," SAE Technical Paper 910454 (1991).

[31] Zhang, Y., Perez, J. M., Pei, P.and Hsu, S. M., "A New Method to Evaluate the Deposit Forming Tendencies of Lubricants by Differential Scanning Calorimetry," Lubr.Eng, Vol.48, No.3, 1992, p.215-220, (1992).

[32] Zhang, Y., Perez, J. M., Pei, P. and Hsu, S. M., "The Deposit Forming Tendencies of Diesel Engine Oils - Correlation Between the TwoPeak Method and Engine Tests," Lubr.Eng.,Vol.48, No.3, 1992, p221-226.

[33] Perez, J. M., Pei, P., Zhang, Y.and Hsu, S. M., "Diesel Deposit Forming Tendencies of Lubricants by Microanalysis Methods," SAE Technical Paper 910750 (1991).

[34] Lahijani, J., Lockwood, F. E. and Klaus, E. E., "The Influence of Metals on Sludge Formation," ASLE Trans., 25(1),1982, p. 25-32.

[35] Klaus, E. E., Shah, P. and Krishnamachar, V., "Development and Use of the Microoxidation Test with Crankcase Oils," NBS SP Publ. No.674, 1982, p.155-168.

Poonam Aggarwal [1], and David Dollimore [2]

THERMAL STUDIES ON THE COMBUSTION PROCESS OF TREATED AND UNTREATED CELLULOSE

REFERENCE: Aggarwal, P. and Dollimore, D., **"Thermal Studies on the Combustion Process of Treated and Untreated Cellulose,"** Oxidative Behavior of Materials by Thermal Analytical Techniques, ASTM STP 1326, A. T. Riga and G. H. Patterson, Eds., American Society for Testing and Materials, 1997.

ABSTRACT: The degradation of cellulose and cellulose samples impregnated with certain salts has been investigated by thermal analysis techniques. The behavior in air and in nitrogen is noted, for both the treated and untreated samples. A comparison is made between the samples using the α_r- α_s plot, which can be used as a tool to examine the solid state reactivities of the materials being compared. The method relies on the calculation of the mass loss data from the thermogravimetry unit being converted into α, the fraction decomposed. This indicated the reactivity of the treated samples to be greater than that for α-cellulose in nitrogen. In air, the sample treated with ammonium phosphate appeared less reactive than α-cellulose.

KEYWORDS: Retardants, thermogravimetry (TG), cellulose.

[1] Department Of Chemistry, University Of Toledo,
Toledo, OH 43606.
[2] Department Of Chemistry and College Of Pharmacy, University Of Toledo,
Toledo, OH 43606.

INTRODUCTION:

Cotton cellulose is a pure glucose polysaccharide, $(C_6H_{10}O_5)n$, and has been used as a starting material for practically all chemical work on cellulose [1]. It contains less than 0.05% ash and is a standard by which other celluloses are judged. Treating cotton cellulose with cold NaOH produces a residue referred to as α-cellulose. The cellulose studied in this investigation is an α-cellulose. α-cellulose was examined by thermal analysis and the degradation it undergoes was studied. It was impregnated with certain flame retardants and the thermal pathway followed by the so formed modified celluloses were also studied to gain information on the effectiveness of the treatment with retardant materials.

The subject of the combustion process of treated and untreated cellulose is associated always with attempts at imparting a degree of fire retardancy to the material. This can be achieved by simply impregnating the material with suitable additives [2-6]. Two kinds of impregnations can easily be recognized, physically adsorbed species present at the surface which can easily be washed away, and chemically bonded species which impart a degree of permanency to the flame resistance of the cellulose fibers. In this latter case the fibers can be laundered, for example and the fire resistance would remain unimpaired. The normal pattern of cellulose combustion is first to produce gaseous products by a process of degradation [7]. In an atmosphere of nitrogen this is an endothermic event. In air the evolved gases catch fire leading to an overall exothermic event [8]. The product at this stage is a carbon or carbonaceous residue, which represents the end product of heating in an inert atmosphere. In air however the carbon itself can undergo combustion to completely gasify with the production mainly of carbon monoxide and carbon dioxide [9]. This is described as glowing combustion. The process is often self sustaining and the exotherm indicating the occurrence of the event may show a 'shoulder' on the DTA trace and it may be better to portray the combustion against time rather than against temperature on any thermal analysis plot. The impregnation with chlorides is an example of fire redardancy caused by physical adsorption of the chloride species. The treatment with phosphorous compounds, e.g. ammonium phosphate, leads to chemical bonding, producing a more permanent effect on the combustion process [10-12]. Unfortunately fire retardancy has been associated with a greater degree of carbon residue being formed. This can be measured by thermogravimetry (TG) plots in nitrogen or inferred by noting the mass remaining between the onset of gaseous combustion and glowing combustion. This has the effect in real fires of producing a greater amount of particulate material and instead of the actual fire being the hazard, the hazard of particulate emission results in a larger danger from suffocation.

In the present study, the extent of particulate material produced by modifications is noted together with other measurements of flame retardancy derived from thermal analysis measurements.

MATERIALS AND METHODS:

α -cellulose was obtained from Scientific Polymers Products. The α cellulose that was used for this work was 99.5 % pure. The TG experiments were conducted using a simultaneous TG -DTA unit from TA instruments. The samples were heated both in an atmosphere of flowing dry air and dry nitrogen, at a heating rate of 10 °C min^{-1}. The sample size used varied from 5-6 mgs. Platinum crucibles were used with empty sample pans used as reference. Repeat experiments with alumina crucibles gave the same results. The salts used for flame redardant treatment were ammonium phosphate, sodium chloride, and potassium chloride. The salts were impregnated on a 2% basis, since this concentration seemed to be typical of work conducted by other researchers. A calculated mass of each salt was dissolved in 50 ml of distilled water. The solution was heated to boiling and then 10 g of α-cellulose was added and the mixture stirred. This was then filtered, dried at 100 °C overnight. For the reference material, pure cellulose was also treated in the same way.

RESULTS AND DISCUSSION:

The TG plot of α-cellulose in nitrogen is shown in Figure 1, showing degradation in a single step to carbon. The DTG data is also shown in this figure and the DTA plot shows a single endothermic peak, the carbon residue amounts to 18%. Thermal analysis data in an atmosphere of nitrogen for the salt treated samples showed a similar overall pattern; all were endothermic, but the initial temperatures, the peak temperatures on the DTG and the residue varied. As an example, the TG-DTG data in nitrogen, for the NaCl treated sample is shown in Figure 2. A summary of the results for the other treated materials is given in Table 1.

TABLE 1: TG data for degradation of treated and untreated cellulose in an atmosphere of dry nitrogen.

Name	% Carbon	Temperature DTG °C			Temperature DTA °C		
		Ti	Tp	Tf	Ti	Tp	Tf
α-cellulose	16.8	220	358	395	296	361	397
Treated with (NH$_4$)$_3$PO$_4$	28.7	195	296	320	205	294	320
Treated with KCl	26.6	220	340	375	274	336	365
Treated with NaCl	20.2	217	343	386	273	340	379

Note : The % carbon is the residue left at 430 °C calculated on the basis of the dried mass of α-cellulose and its modifications. This varies by a small amount from the direct reading on the TG plot.
T_i is the initial temperature, T_p the peak temperature and T_f is the final temperature of the DTG curve.

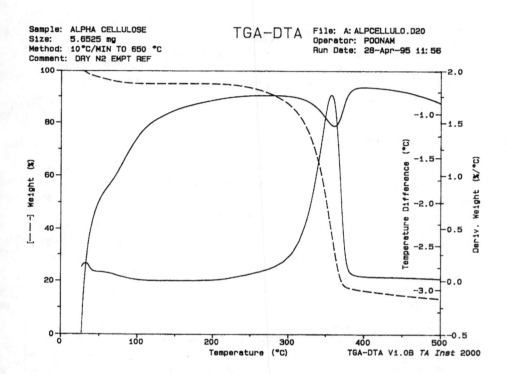

FIG.1   TG plot of α-cellulose in nitrogen.

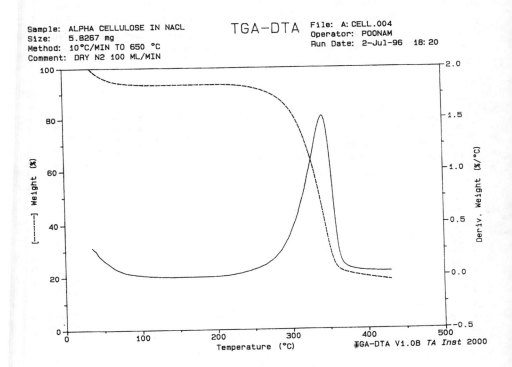

FIG.2   TG-DTG data in nitrogen for NaCl treated sample.

The TG data for combustion of the cellulose samples shows evidence of self sustaining exothermic events. This is evident from the TG plot of the % weight against temperature for the NaCl treated sample. In this region the sample has caught fire (a self-sustaining combustion process). The presentation can however be resolved by presenting the data as % weight against time (see Figure 3). The sample displays two regions of mass loss. This was also confirmed by the DTA plot (Figure 4). In some cases initial mass loss was associated with a double overlapping DTA plot (the original sample and the KCl treated sample), whilst in the case of KCl the slower period of mass loss between the gaseous combustion region and the self sustaining combustion process was also associated with an exothermic peak (Figure 5). The cellulose sample treated with the $(NH_4)_3PO_4$ showed no sign of a self-sustaining exotherm (see Figure 6) which would emphasize its use as a fire retardant. The data all these combustion processes as drawn from the TG, DTG and DTA data is presented in Table 2.

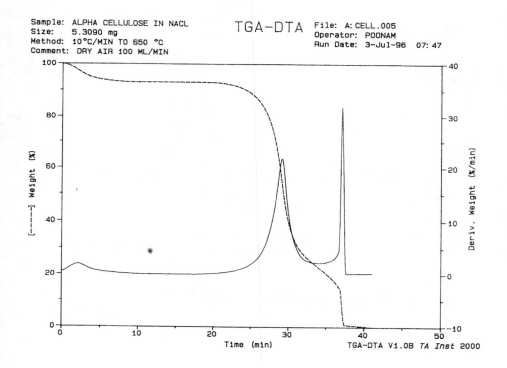

FIG.3 TG data as % weight against time.

TABLE 2: TG data for pyrolysis of treated and untreated cellulose

Sample	alpha cellulose	$(NH_4)_3PO_4$	KCL	NaCl
TG/DTG % mass loss, gaseous combustion	71.5 %	68 %	72 %	63 %
% mass loss, glowing combustion	17.9 %	31 %	23 %	17 %
% mass loss, self sustaining glowing combustion	8 %	0 %	4 %	18 %
Initial Temperature	247 °C	214 °C	242 °C	241 °C
Temperature at the end of % mass loss, gaseous combustion	349 °C	285 °C	354 °C	331 °C
Temperature at the end of % mass loss, glowing combustion	470 °C	577 °C	428 °C	389 °C
Reaction temperature interval for gaseous combustion	102 °C	72 °C	112 °C	90 °C
Reaction temperature interval for glowing combustion	104 °C	292 °C	74 °C	58 °C

Note: The % mass has been amended in all cases to make the data refer to the dried mass of the α-cellulose and its modifications as noted in Table 1.

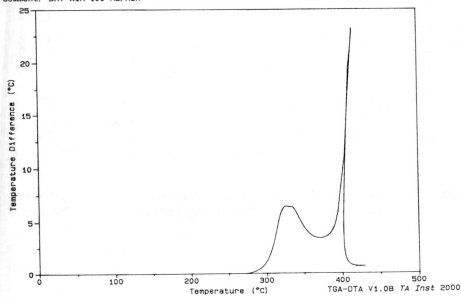

Sample: ALPHA CELLULOSE IN NACL   TGA−DTA   File: A:CELL.005
Size: 5.3090 mg   Operator: POONAM
Method: 10°C/MIN TO 650 °C   Run Date: 3−Jul−96   07:47
Comment: DRY AIR 100 ML/MIN

TGA-DTA V1.0B *TA Inst* 2000

FIG.4   DTA plot confirming two regions of mass loss.

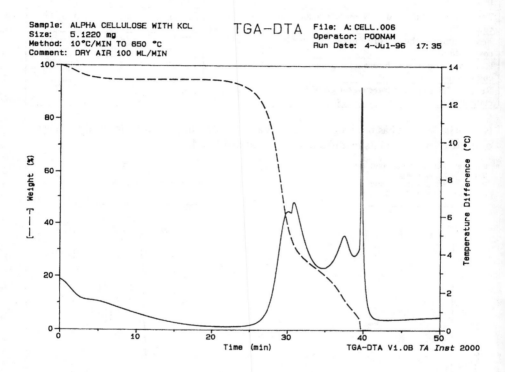

```
Sample:  ALPHA CELLULOSE WITH KCL        TGA-DTA    File:  A: CELL.006
Size:    5.1220 mg                                  Operator:  POONAM
Method:  10°C/MIN TO 650 °C                         Run Date:  4-Jul-96   17: 35
Comment: DRY AIR 100 ML/MIN
```

FIG.5   In the case of KCl, the slower period of mass loss between the gaseous combustion region and the self-sustaining process was associated with an exothermic peak.

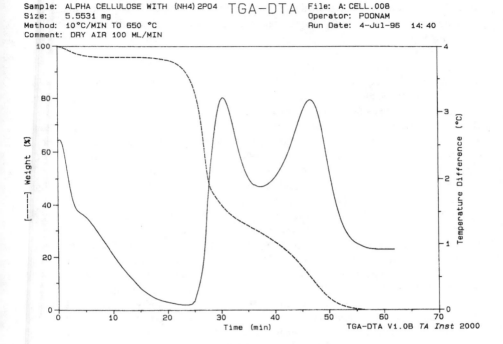

Sample: ALPHA CELLULOSE WITH (NH4) 2PO4   TGA—DTA   File: A: CELL.008
Size: 5.5531 mg   Operator: POONAM
Method: 10 °C/MIN TO 650 °C   Run Date: 4-Jul-96 14: 40
Comment: DRY AIR 100 ML/MIN

FIG.6   Cellulose sample treated with $(NH_4)_3PO_4$ showed no sign of a self-sustaining exotherm.

The data could be interpreted in terms of traditional kinetic data evaluation which would involve an assessment of the kinetic mechanism, and the Arrhenius parameters [13]. However there is a comparative assessment of reactivity available developed initially to show the effect of milling on the reactivity of crystals [14]. In this method one of the samples under study is chosen as the reference and the others are called sample. For a specified reaction sequence the value of the fraction decomposed at any particular temperature for the sample and the reference material are noted. α is determined by the following equation:

$$\alpha = \frac{w_i - w}{w_i - w}$$

where w= % weight of substance, w_i = % weight of substance at initial time and w_f = % weight of substance at final time.

The data can be presented as a plot of α_r- α_s. If the reactivity of the sample is the same as that of the reference then a single linear plot will result. If however the reference or the sample is more reactive then a deviation form the reference line in the upper or lower direction is seen. This gives information about which material has higher reactivity. In this case the data for the treated samples (called α_s) was compared with that of the original unmodified α-cellulose (called α_r) for the TG data in nitrogen is shown in Figure 7. In this plot all the modified samples showed higher reactivity than the unmodified cellulose. The ammonium phosphate treated sample clearly shows a greater reactivity as compared to the other samples. The α_r- α_s plot for the combustion of the treated and unmodified celluloses is given in Figure 8. This shows the same sequence of events for the gaseous combustion region, but the glowing combustion sequence indicates that the sample treated with ammonium phosphate less reactive than the original cellulose. Tables of these for α_r- α_s plots indicating the α_r/ α_s ratio data for nitrogen are given in Table 3 for α_r values of 0.1 and 0.9. Similar data for the combustion data is provided in Table 4. With regard to fire retardancy, (this refers to the nature of the process of combustion) the data from the simultaneous thermal analysis is conflicting. This general comment agrees with earlier investigations by Hoath [15]. Features of the available thermal analysis data pertinent to flammability can now be discussed.

(1) The onset of decomposition / combustion.

It should be noted that the order of the initial temperature, of degradation in nitrogen is (from Table 1) α-cellulose $\equiv$ KCl> NaCl> $(NH_4)_3PO_4$,

(where the symbols refer to the kind of treatment received by the cellulose). In the combustion data the order is (from Table 2, % mass loss)

α-cellulose > KCl $\equiv$ NaCl> $(NH_4)_3PO_4$,

This taken alone would suggest that none of these additives are effective fire retardants. The T_i and T_f data would also corroborate these findings (see Table 1).

(2) Other temperature dependent findings

The direct data on the combustion process (see Table 2) indicates that treatment of the cellulose does not significantly increase the temperature at the end of the gaseous combustion period or at the end of the glowing combustion period. The exception is the ammonium phosphate treated sample, where the end of the glowing combustion period is at a much higher temperature than for any of the other samples. The temperature interval for gaseous combustion is also reduced. When the absence of a self sustaining combustion process is taken into consideration for the $(NH_4)_3PO_4$ treated sample, the evidence points to the $(NH_4)_3PO_4$ treatment as an effective fire retardant treatment.

(3) The total mass of carbon produced in the thermal analysis under an atmosphere of nitrogen.

Table 1 shows that the total amount of carbon produced is in the order

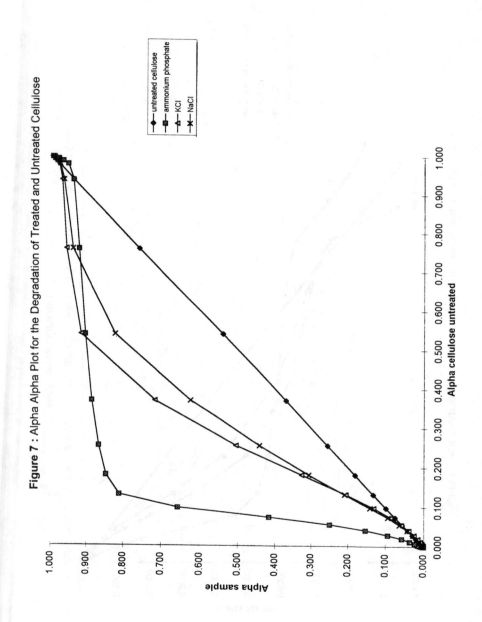

Figure 7 : Alpha Alpha Plot for the Degradation of Treated and Untreated Cellulose

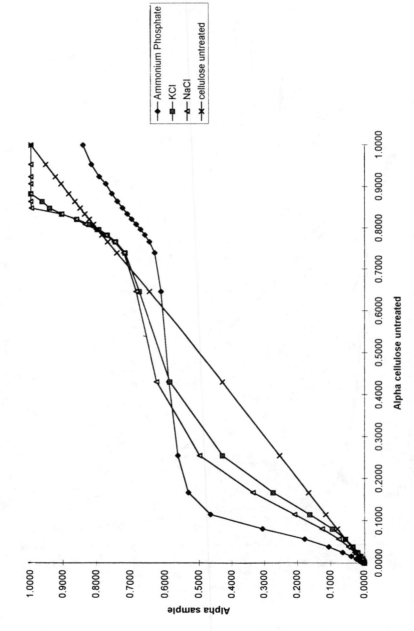

Figure 8 : Alpha -alpha plot for the combustion of treated and untreated cellulose

TABLE 3:

$\alpha_{T/}\,\alpha_s$ value for treated and untreated samples at $\alpha_{T/}\,\alpha_s$ values of 0.1 and 0.9 respectively for degradation in an atmosphere of nitrogen.

Sample	$\alpha_r = 0.1$	$\alpha_r = 0.9$
$\alpha_{T/}\,\alpha_s$ $(NH_4)_3PO_4$	0.25	1.17
$\alpha_{T/}\,\alpha_s$ KCl	0.62	0.90
$\alpha_{T/}\,\alpha_s$ NaCl	0.50	0.90

TABLE 4:

$\alpha_{T/}\,\alpha_s$ value for treated and untreated samples at $\alpha_{T/}\,\alpha_s$ values of 0.1 and 0.9 respectively for the combustion.

Sample	$\alpha_r = 0.1$	$\alpha_r = 0.9$
$\alpha_{T/}\,\alpha_s$ $(NH_4)_3PO_4$	0.15	0wazzu .99
$\alpha_{T/}\,\alpha_s$ KCl	0.80	0.97
$\alpha_{T/}\,\alpha_s$ NaCl	0.71	0.0.98

$(NH_4)_3PO_4 > KCl > NaCl > \alpha$-cellulose.

The data for combustion showing the effect of gaseous combustion (see % mass loss in Table 2) shows % carbon present before onset of glowing combustion as

$NaCl > (NH_4)_3PO_4 > \alpha$-cellulose $\equiv KCl$

The evidence here is not so clear as the carbon formed can undergo glowing combustion before the gaseous pyrolysis is complete. The tendency is to relate fire is retardance efficiency in terms of increased production of particulate material in the initial region of combustion which would mean that the addition of these fire retardant materials to the cellulose makes a contribution to fire retardancy. The remaining carbon residue is lost in the glowing combustion zone (see Table 2). In a real fire situation much of this particulate material, consumed in the glowing combustion zone, is in fact lost and makes a significant perhaps major contribution to the fumes and smoke which arise from combustion. However this glowing combustion process can be divided into two zones, one in which heat is supplied to the system and a final zone representing a self sustaining combustion (in terms of mass loss) which by inspection of Table 2 is in the order $NaCl > \alpha$-cellulose $> KCl > (NH_4)_3PO_4$

It should be noted that in the case of the $(NH_4)_3PO_4$ treated sample there is no zone of self sustained combustion and this fact can be used to support the use of $(NH_4)_3PO_4$ as imparting fire retardant properties on the α-cellulose. NaCl treatment on the other hand increases the mass loss due to self - sustained combustion.

(4) The α_r- α_s method of assessing reactivity in combustion.

From the data collected in nitrogen, it can be seen that in the region corresponding to gaseous combustion the reactivity of all the treated samples is greater than the untreated cellulose (see Figure 7 and Table 3). For the combustion process the treated samples are initially more reactive than the untreated cellulose in the zone of gaseous combustion. However, in the region of glowing combustion the $(NH_4)_3PO_4$ treated cellulose is less reactive than the untreated one, whilst the chloride treated cellulose remains more reactive (see Figure 8, Table 4).

CONCLUSIONS:

The above data reveals that fire retardants should be considered for effectiveness with regard to three zones of combustion:

1. Gaseous combustion
2. Glowing combustion in which when the source of heat is removed the fire will eventually stop and
3. Self - sustained glowing combustion.

It is suggested that fire retardants effective in any one zone may not be so effective in the other two zones. The $(NH_4)_3PO_4$ is particularly effective when applied to α-cellulose in suppressing the self - sustained zone of combustion.

REFERENCES:

1. Pigman, W., ed., The Carbohydrates, Chemistry, Biochemistry, Physiology, p. 21. Academic Press Inc., New York, 1957.
2. Wanna, J.T., and Powell, J.E., Thermochimica Acta, Vol., 226, 1993, 257-263.
3. Sekiguchi, Y., Shafizadeh, F., Journal of Applied Polymer Science, Vol., 29, 1984, 1267.
4. Khattab, M.A., Price, D.,and Horrocks, A.R., Journal of Applied Polymer Science, Vol., 41, 1990, 3069.
5. Horrocks, A.R., Davies, D., and Greenhalgh, M., Fire and Materials, Vol 9, 1985, 57.
6. Hirata, T., and Nishimoto, T., Thermochimica Acta, Vol., 193, 1991, 99-106.
7. Shafizadeh, F., Advances in Carbohydrate Chemistry, Vol 23, 1968, 419.
8. Fardell, D.J., Lukas, Lukas, C., Chemistry in Britain, Vol 23, 1987, 221.
9. Shafizadeh, F., Lai, J. Z., Journal of Organic Chemistry, Vol 37, 1972, 278.
10. Lamby, E. J., Spiro, C.L., McKee, D.W., Carbon Vol 22, 1984, 285.
11. Jain, R.K., Lal, K., Bhatnagar, H. L., Journal of Polymer Science, Vol 30, 1985, 897.
12. Lyons, J. W., Journal of Fire and Flammability, Vol1, 1970, 302.
13. Dollimore, D., Hoath J.M., Thermochimica Acta, Vol., 121, 1987, 273-282.
14. Heda, P. K., Dollimore, D., Alexander, K. S., Chen, D., Law, E., Bicknell, P., Thermochimica Acta., Vol 255, 1995, 255.
15. Dollimore, D., and Hoath J.M., Thermochimica Acta, Vol., 45, 1981, 87-102.

Lecon Woo [1], Samuel Y. Ding [2],
Michael T. K. Ling [2] and Stanley P. Westphal [2]

**FURTHER STUDIES ON OXIDATIVE INDUCTION TEST
ON MEDICAL POLYMERS**

REFERENCE: Woo, L., Ding, S., Ling, M. T. K., and Westphal, S. P., **"Further
Studies on Oxidative Induction Test on Medical Polymers,"** Oxidative Behavior of
Materials by Thermal Analytical Techniques, ASTM STP 1326, A. T. Riga and G. H.
Patterson, Eds., American Society for Testing and Materials, 1997.

ABSTRACT: We have published previously initial results on
the application of the oxidative induction test to a wide
variety of medical polymers. This is a continuation and
further elaboration on the earlier study.

For medical flexible polyvinyl chloride (PVC) compounds,
the traditional measure of degree of degradation by color
formation was found to correlate to measured oxidative
induction times. Furthermore, three distinct regimes were
also detected. In the initial phase of degradation, little
color formation was detected for significant decreases in the
induction time. This is followed by a nearly linear regime.
Finally in the terminal regime, where the stabilizer were
nearly exhausted the color increased exponentially. This
observation was interpreted by the established mechanism of
PVC degradation.

In another series with polyolefins, several subtle
phenomena were observed that may have utility in either
optimizing the stabilization package or predicting long term
shelf life. It was found that significant antioxidant losses
occurred at typical experimental temperatures above 200°C.
Much better shelf life predictions can be obtained by
collecting data at lower temperatures. And the inflection
point on the activation energy plot corresponded to the
antioxidant volatilization temperature.

KEYWORDS: Oxidative induction test, OIT, PVC, polyolefins.

1 Baxter distinguished scientist, Medical Materials
Technology center, Corporate Research and Technical Services,
Baxter Healthcare, Round Lake, IL 60048.

2 Senior engineering specialists, Corporate Research and
Technical Services, Baxter Healthcare, Round Lake, IL 60048.

INTRODUCTION:

We have previously studied the utility of the oxidative induction test applied to many of the medical polymers [1]. It was found that in numerous cases, if a clear induction point can be found, important insights can be generated on polymer stability, antioxidant formulation effectiveness, as well as in product and process performances. In this study, several systems are examined in further detail to gain ' additional information about optimization of the product or process.

EXPERIMENTAL PROCEDURE:

ASTM Procedure D3895-80 is followed in the main, except air is used instead the pure oxygen, and both the isothermal oxidative induction time (OIT) and temperature scanning induction temperature were used. The oxidative induction test was conducted on a Dupont 1090 thermal analyzer with 910 differential scanning calorimetry (DSC) cell. Normally this test has two modes of measurement, that is, oxidative induction temperature and oxidative induction time. Oxidative induction temperature measures the onset of auto-oxidative reaction while the temperature is scanning at a preset rate. For this mode of testing, usually a thin and flat specimen, typically about 2 mg or less, was prepared and placed in an open aluminum sample pan and secured on the thermoelectric disk of the DSC cell. The sample was then scanned at a rate of 20°C/min from ambient to 300°C or higher in an air purging stream of 100 ml/min. The second method, the oxidative induction time, is a relative measure of the degree or level of stabilization of the material tested. The specimen preparation is the same as in the continuous temperature scanning method, except for scanning in a nitrogen gas environment to the preset isothermal testing temperature. Once temperature equilibrium has been established, the controller automatically switches purge gas to air or oxygen at the same purging rate. The changeover point to air or oxygen purge is taken as the zero time of the experiment. The oxidative induction stability of tested samples is assessed by monitoring an abrupt exotherm or endotherm departure from the baseline as indicated by Figure 1 [2].

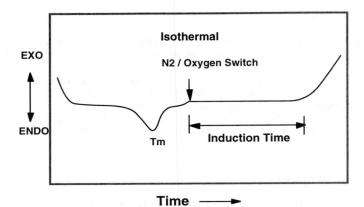

Time ——▶

FIG.1--OIT tracing from Differential Scanning Calorimeter.

Often, if the oxidative reaction follows simple kinetics order, induction time measured at various temperatures can be used to construct an Arrhenius plot, expressed as $\log(OIT^{-1})$ versus T^{-1}, to obtain information on the oxidation reaction kinetics. Mathematically, the rate constant K, which is proportional to OIT^{-1}, may be expressed in the Arrhenius form as:

$$K=K_o e^{(-\Delta E/RT)} \tag{1}$$

where K_o is the pre-exponential factor, ΔE the activation energy of the reaction, R is the gas constant, and T is the absolute temperature in Kelvin. The slope of the log K versus T^{-1} plot is then the activation energy divided by R. Specimens less than 100 μm in thickness were used through out this study to ensure homogeneity. Inhomogeneity could result in multiple oxidative transitions thus, it should be avoided. Inhomogeneity usually arises from skin and core of molded sections, spatial variations in composition or thermal histories. With thin sections and small sample sizes, all of the inhomogeneities are likely to be resolved, and differentiation of skin and core, thermal histories made possible and meaningful.

PVC FORMULATIONS

Polyvinyl chloride is a unique polymer providing wide property spans in the medical field by simply adjusting the level of plasticizers in the formulation. In selecting different base polymers, the intrinsic stability of the neat polymer is an important consideration. The intrinsic stability arises from the perfection of the polymer main-chain, head to head additions, degrees of unsaturation and is mainly a function   of the polymerization process conditions during manufacturing. It is important for medical applications to select resins with minimum chain imperfections which are more resistant to degradation and the formation of impurities and extractables.

The degradation of the PVC is primarily a cationic dehydrochlorination chain reaction [Eq. 2], and most stabilizing schemes are directed toward the elimination of labile reaction sites and the sequestration of the hydrogen chloride as soon as it is produced to avoid further catalyzing neighboring groups.

$$[- C- C- C- C-]_n \longrightarrow [- C= C- C=C-]_n + 2 n \, H \, Cl \qquad (2)$$
$$\quad\;\; | \qquad | $$
$$\quad\; Cl \quad\;\; Cl$$

We studied the comparison of suspension PVC resins from different suppliers. It was noted that based on various factors, among them, polymerization temperature, size of the reactor and agitation, type of initiator and suspension agents, degree of conversion and whether chain transfer agents are used to control molecular weight, various chain ends and chain imperfections would be created (Figure 2). These factors should lead to measurable differences in the oxidative induction time.

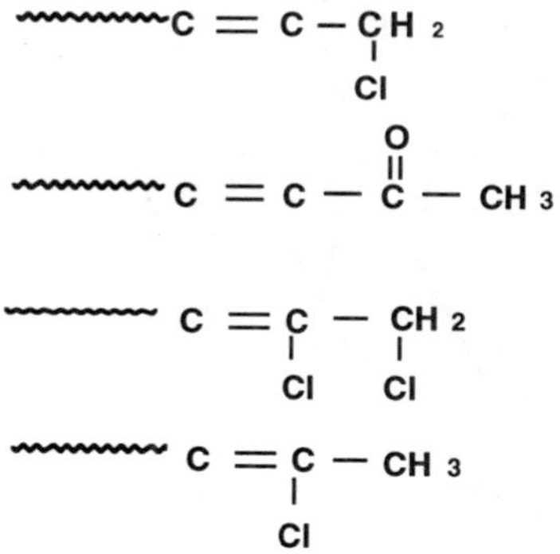

FIG. 2--Suspension PVC chain ends, [3].

We reported in an earlier study on the PVC thermal stability dependence on formulations. OIT was found to exhibit a rather sharp maximum with respect to the primary stabilizer, calcium zinc stearate at about 0.13 weight %

(Figure 3), while an extremely linear relationship was found for the secondary stabilizer, an epoxidized oil. In summary, the PVC stability function spanned a three dimensional design space schematically depicted in Figure 4. The utility of graphs like Figure 4 in polymer stabilization system development is readily apparent.

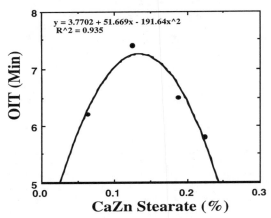

$$y = 3.7702 + 51.669x - 191.64x^2$$
$$R^2 = 0.935$$

FIG. 3-- PVC OIT dependence on CaZn stearate.

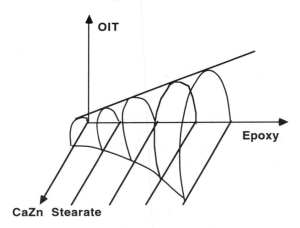

FIG. 4-- 3-Dimensional schematic of PVC stability.

Base (neat) PVC polymer powder does not exhibit a sharp OIT onset. However, a semi-micro scale solution formulation of PVC, calcium zinc stearate, and plasticizers with epoxydized oil, was found to simulate PVC compounds produced on large scale mixers and melt extrusion lines. Briefly, commercial formulations were replicated in a fractional gram scale, except tetrahydrofuran was used to solution blend all ingredients. After the solvent has been thoroughly removed in a vacuum oven, the resulting film was made more homogenous by repeated folding and compression molding at a relatively low temperature of about 180°C. The OIT data at 230 °C are shown in Figure 5 with the standard deviation shown as the error bars.

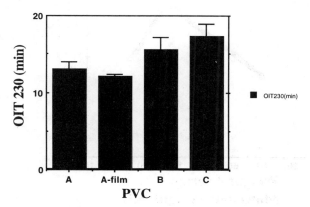

FIG. 5--PVC resin stability comparison.

Figure 5 also compares the formulation stability from resin supplier A prepared by two methods, the semi-micro laboratory method and a large scale extruded film. The slightly extended induction time from the laboratory sample clearly indicated the less severe thermal and shear histories compared with extrusion. Moreover, the significantly greater standard deviation also reflected less homogeneity. In a similar way, the resins from suppliers B and C are significantly more stable inherently without going through a large scale compounding and extrusion step.

In a separate experiment suspension PVC powders from Supplier A with different molecular weights were similarly compared for their inherent stability, and their OIT at 230°C plotted against the molecular weight, or the number of chain ends per unit volume. When the results in Figure 6 were examined, it showed that the sample at about 110 KD molecular weight exhibited an anomalous high stability as one would expect from the concentration of chain-end imperfections. However, an inquiry into the PVC manufacturing process revealed that for this polymer, the molecular weight was controlled by the addition of a chain transfer agent. Since the chain transfer agent would invariably "cap" the reactive chain-ends, enhanced inherent stability was achieved. This example clearly revealed that when a series of polymers of similar compositions were compared under identical

conditions, very subtle, often chemically difficult to
determine structural differences can be obtained by OIT.

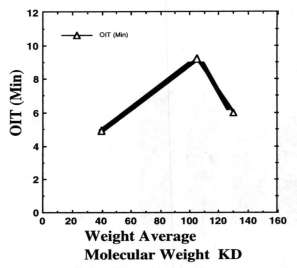

FIG. 6--Supplier A PVC OIT versus weight average
molecular weight.

To further demonstrate the utility of the test, PVC
films subjected to a standard heat aging test were used. In
the heat aging test, film samples were conditioned in a
circulating hot air oven at 190°C for various times and the
resultant color formation is a measure of relative processing
stability. In this case ASTM test of yellowness index [4] was
used for the color determination. The basis of color came
from the cationic dehydrohalogenation degradation mechanism
for PVC. As hydrogen chloride is liberated from the PVC
matrix, it coordinates onto a neighboring tertiary carbon
(most reactive), forming a very stable allylic carbonium ion,
before eliminating another HCl molecule to propagate the
reaction. In so doing, a series of conjugated dienes (polyenes) were formed. Since the π electrons on these conjugated
dienes can freely move over the entire length of the
conjugated diene, an one dimensional electron well resulted.
As the diene increases in length, the energy levels and the
absorption spectra of the free electron begin to move from UV
toward the visible wavelengths. While the absorption spectra
increase in intensity from the short wave length direction,
the originally clear PVC film starts to appear yellow to dark
yellow, orange, red, and finally black in the degradation
progression.

When the yellowness index of these films were compared with OIT measured at 230°C, a very good correspondence was obtained (Figure 7).

Clearly one can uniquely determine the degree of the degradation compared to the color formation as evidenced by the fit of the data, subtle consequences from the mechanisms of the degradation are also evident. Figure 7 clearly indicates that the degradation of the PVC proceeds in three distinct phases.

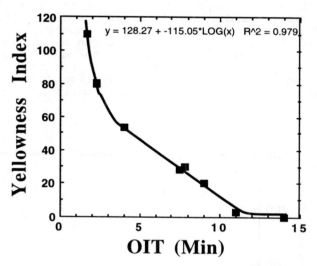

FIG. 7-- PVC OIT, color correlation.

During the initial phase, significant induction time was lost without detectable color shift. The second phase followed a more or less linear dependence of OIT with color. And in the final phase, as the stabilizer becoming exhausted and numerous poly-ene sequences simultaneously moving in from the ultraviolet, the color intensity becomes exponential before total destruction. The first and initial phase deserves further comment. If, in the absence of significant and detectable color, extensive degradation reaction can take place as measured by OIT, the consequence on chemical extractable must be extensive. This may offer a rapid and effective screening method on chemical and leachable testing.

In terms of the color formation, a simplified model based on the polyene sequences can be formulated to explain this three-step behavior in color formation. Assuming that the polyene formation follows an initiation, propagation and termination mechanism, the induction phase in color formation at the beginning of the degradation be demonstrated. As the stabilizer gets consumed and the corresponding oxidative induction time decreases, the length of the polyene sequences slowly grows. Since polyenes with length shorter than four units only absorb light in the ultraviolet, there will be no visible color formation at the beginning of the degradation. The induction phase ends when polyenes, with length greater than five units begin to accumulate, and the yellowness index increases after significant degradation has taken place.

After the induction phase, the linear increase in yellowness index can be produced with a first order reaction for the initiation and propagation steps and a second order reaction for the termination step. The exponential increase in yellowness index at the final stage can be accurately modeled if the autocatalytic effect by the hydrochloric acid generated during the polyene formation is taken into account. Figure 8 shows the simulated color formation during degradation based on this simplified mechanism.

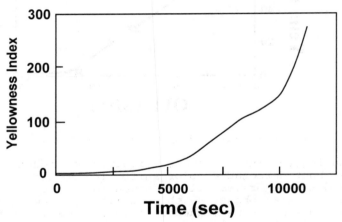

FIG. 8-- Model simulation of yellowness index versus time.

POLYOLEFINS

Due to their inherent cleanliness and cost/performance, polyolefins are desirable medical polymers. Since these are the polymers used when the original OIT test was established by researchers in the Bell Telephone Laboratories, a more detailed examination is taken. Gugamus[5] discussed the lack of correlation between OIT measurements on polypropylene and long term oven aging results. The oven aging results were obtained by onsets of embrittlement in up to 14 weeks of oven aging at 135°C. However, a close examination of the data

(Figure 9) indicated that only a single data point was out
of place. With the offending data removed, the data fit
improved from a correlation coefficient of 0.49 to 0.69
(Figure 9a), a remarkable improvement indeed. Since the
embrittlement assessment are known to be subjected to large
errors, most of the data scatter in Figure 9 was expected to
have come from the oven aging test.

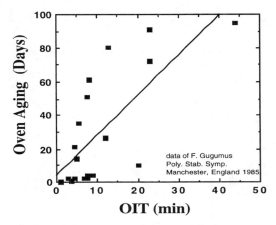

FIG. 9-- Polyolefin OIT oven stability comparison

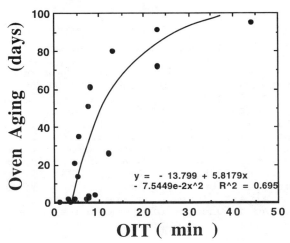

FIG.9a--Replot of Figure 9 with one data point
removed.

This, combined with the very rapid nature(minutes) of
the OIT test, re-affirms the value of the OIT as a early
predictor of the long term aging results.

When the induction time is plotted in the Arrhenius
fashion, a curious inflection in the activation enthalpy is
frequently seen at relatively high temperatures. On the low
temperature side of the line, a normal activation enthalpy is
measured. On the high temperature high of the inflection, an
anomalous higher activation enthalpy is measured. We
speculated that the increase in apparent degradation rate is
the volatilization of antioxidants at these high
temperatures. For example, a highly stabilized structural
polypropylene, the inflection occurred at 2.06 on the 1000/K
scale, or about 210°C (Figure 10).

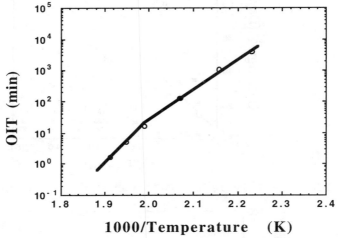

FIG. 10-- Polypropylene OIT versus 1000/K

However, since the stabilizer package is not known, we can
only suspect but not attribute the observation to the lost
of antioxidants. In a similar series where we have
specifically compounded 0.1 wt.% of Irganox-1010 [6]
antioxidant in high density polyethylene (HDPE), an
inflection point is clearly visible at about 2.015 x10-3 or
200°C. Below 200°C, an activation enthalpy of about 80 KJ/
mole is seen, while above 200°C, a much higher activation
enthalpy of about 250 KJ/mole was seen. Bair[7] had published
rate of volati-lization of Irganox-1010 determined by thermal
gravimetric analysis. Indeed, if one replots the data from
the reference as a function of temperature, a strong upturn
is seen at about 200°C (Figure 11).

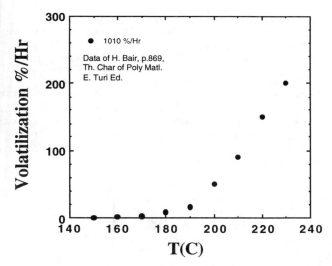

FIG. 11--Phenonic antioxidant volatilization

Based on this we can conclude that these high temperature OIT inflection points are most likely caused by antioxidant volatilization. Following this line of reasoning, OIT data measured below the inflection, those free from the loss of antioxidants should be used for lower temperature property predictions.

In a similar way, a polypropylene sample with Ethyl-330 antioxidant was found to have a deflection at about 190°C.

POLYESTER ELASTOMER

Polyester thermoplastic elastomers (TPE) based on polybutylene terephthalate (PBT) hard segment [8] and tetramethylene ether (PTMO) soft segments constitute an important class of medical elastomers because of their wide property range, solvent bonding capability, oxidative stability and processing ease. In a series of experiments film samples of 40D hardness were prepared on a 38mm laboratory extruder at similar processing conditions but at varying thickness. Figure 12 presents the OIT activation energy plot of two of the films. Both the 100 μm(4 mils) and 200 μm (8 mils) film exhibited identical activation energies, indicative of identical chemistry. A two layer composite of the thinner film also exhibited identical OIT of the thinner film. This indicates that the sample thickness of the OIT sample is not a primary variable for the induction times measured. However, between the two sets of data, there was a four fold difference. An inquiry into the details of the film extrusion process revealed that the thinner film was processed at slightly higher temperatures (193°C vs. 175°C). However, temperature alone could not account for the magnitude of the observed OIT discrepancy. It was concluded that the residence time in processing the thinner film must have been significantly longer than the 200 μm film. In other words, Figure 12 maps out a time temperature envelop for processing this particular polymer without significantly depleting the antioxidant package.

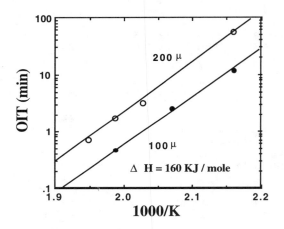

FIG. 12--Polyester TPE Processing

SUMMARY:

      In this continuation of an earlier study, we have examined in detail the application of the oxidative induction test to more polymer systems in the medical industry. In the case of flexible polyvinyl chloride (PVC), subtle effects of chain perfection were detected by using identical formulations. For polyolefins, it was found that a high temperature OIT acceleration resulted from the volatilization of antioxidants. And in comparing OIT's from a polyester thermoplastic elastomer, the processing temperature and residence time combination was identified as the principal variable in minimizing processing degradations.

REFERENCES

(1)   Woo,L., Ling, Michael T. K., and Chan, E. K., Journal of Vinyl Technology, 13, No.4, p. 199 (1991).

(2)   ASTM   Standard D3895.
(3)   Hjertberg, T., and Sorvik, E. M., <u>J. Makromol. Sci.Chem</u>.
<u>A17</u>(6), 983, (1982).
(4)   ASTM Standard Vol. 8.01, D1925.
(5)   Gugamus, F., <u>Proce. of Polymer Stabilization Symposium,</u>
<u>Manchester</u>, England, (1985).
(6)   Trade Mark, Ciba Geigy.
(7)   Bair, H. E., <u>Thermal Characterization of Polymeric</u>
<u>Materials</u>, p. 869, E. Turi Ed., Academic Press, New York,
(1981).
(8)   Legge, N. R., Holden, G., and Schroeder, H. E., Eds,
<u>Thermoplastic Elastomers</u>, Hanser Macmillan, New York, (1987).

A. Richard Horrocks,[1] Joseph Mwila,[2] and Mingguang Liu[1]

THE USE OF THERMAL ANALYSIS TO ASSESS OXIDATIVE DAMAGE IN POLYOLEFINS

REFERENCE: Horrocks, A. R., Mwila, J., and Liu, M., **"The Use of Thermal Analysis to Assess Oxidative Damage in Polyolefins,"** Oxidative Behavior of Materials by Thermal Analytical Techniques, ASTM STP 1326, A. T. Riga and G. H. Patterson, Eds., American Society for Testing and Materials, 1997.

ABSTRACT: During the thermal oxidation of polyolefins, exemplified by isotactic polypropylene, few observable physical and chemical changes occur until embrittlement or similar failure. The increasing oxidative and auto-oxidative behavior that accompanies oxidation can be measured as a reducing oxidation induction time using isothermal thermal analysis. Alternatively, dynamic thermal analysis (DSC or TGA) may be used more conveniently to record the shift to lower temperatures of the post-fusion, oxidative exotherm quantified as an onset (T_{on}) temperature. This paper collates data from a number of previous and current studies on the oxidative behavior of oriented polypropylene tapes and filaments exposed at elevated (130°C) temperature in air. During exposure, T_{on} values decrease according to a power law dependence with time and these shifts may be used to assess degrees of oxidation present in aged specimens. The implications of these shifts are discussed in terms of monitoring in-service behavior of exposed polyolefins and their relationship to oxidative and auto-oxidative mechanisms are discussed.

KEYWORDS: thermal analysis, DSC, TGA, polypropylene, oxidation, auto-oxidation, oxidative degradation

---

Polyolefins during both processing and service undergo oxidative attack from oxygen in air and other oxidising species present as trace impurities (for example, heavy metal ions) or in the immediate environment (for example, ozone, oxides of nitrogen, sulphur, and so forth). The sensitizing effect of light and accelerating influence of heat are well known to increase rates of oxidative attack [1-3]. The oxidative mechanisms present in polyolefins are specific to the polymer and oxidising condition. In particular,

---

[1]Professor and research student, respectively, Faculty of Technology, Bolton Institute, Bolton, BL3 5AB, UK.
[2]Manager, Dunlop (Z) Limited, P.O. Box 71650, Ndola, Zambia.

the lability of tertiary hydrogen atoms, in polypropylene, is responsible for its greater sensitivity to thermal oxidation than polyethylene; tertiary hydrogen atoms at branch points in low and linear low density polyethylenes are similarly sensitive. While exact mechanisms of photo- [2] and thermal [3] oxidation of polyolefins are complex and are still not fully understood, there is a desire to measure changes in certain polymer physico-chemical properties which indicate levels of oxidative attack experienced during service and effective remaining useful lifetimes. Presence of oxidative degradation in polyolefins is typically determined by infrared spectrophotometry in which oxidation intermediates and such products as hydroperoxides and carbonyl species are easily identified [1]. However, in polypropylene we have shown that during its thermal oxidation, significant changes in tensile properties may occur with little or no formation of hydroperoxide and carbonyl species until after embrittlement or failure has occurred [4]; this agrees with findings in earlier research [5]. It is difficult, therefore, to monitor levels of oxidative degradation occurring before this impending embrittlement of the polymer and, more importantly, the prediction of remaining effective lifetimes of exposed polypropylene becomes impossible. This and other concurrent research [6-8] suggests that small levels of oxidation within polypropylene and, more recently, polyethylene, may be identified by thermal analysis as changes in the behavior of the postfusion, auto-oxidative exotherms.

   This paper will reexamine earlier published data and, together with more recent as-yet unpublished thermal analytical data, will demonstrate the usefulness of both DSC (and DTA) and TGA in monitoring levels of thermal and photo-oxidation in selected polyolefins, particularly polypropylene.

$$PH \longrightarrow P\cdot + XH$$

$$P\cdot \xrightarrow{O_2} PO_2\cdot$$

$$PO_2\cdot + PH \longrightarrow POOH + P\cdot$$

$$POOH \longrightarrow PO\cdot + OH\cdot$$

$$PH + PO\cdot, OH\cdot \longrightarrow P\cdot + POH, H_2O$$

$$PO\cdot, OH\cdot \longrightarrow \text{further oxidation and chain scission products}$$

Fig. 1 -- Polyolefin oxidation mechanisms [9]

AUTO-OXIDATIVE TRANSITIONS AND THERMAL ANALYSIS

It is generally assumed that the oxidation of polyolefins follows the Bolland and Gee [9] mechanism (Fig. 1). The oxidized polymer segments whether comprising oxygenated side groups, main chain groups, or end groups will be sensitive to auto-oxidation during subsequent thermal treatment as experienced, for example, during subsequent thermal analytical experiments. Dynamic DTA responses (Fig. 2) for an unstabilized , reactor-grade polypropylene polymer and derived, antioxidant stabilized orientated films (draw ratio = 3.8:1) from polymer PP4 (Table 1) in air and nitrogen show that several changes in the postfusion, auto-oxidative exotherm are evident [4].

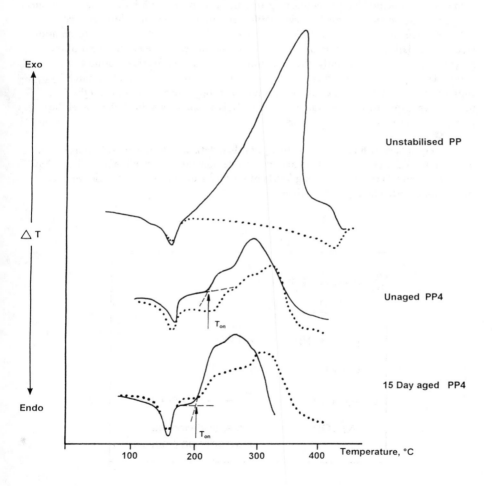

Fig. 2 -- <u>Dynamic DTA responses in air (——) and nitrogen (······) at 20 K min^{-1}</u>

These are summarized below:-

1. Unstablized, reactor-grade polypropylene shows an intense oxidation exotherm only in the presence of oxygen; this is not surprising since this the polymer should contain no oxidative centers.
2. After processing into an orientated film via an additional thermal stage which introduces antioxidant and converts the powder to granular form, the postfusion exotherm is largely auto-oxidative as shown by its intensity under nitrogen.
3. Thermally aging for 15 days in air at 130°C, which is a time just below the embrittlement time of 20 days, changes the overall shape of the exotherm (under both nitrogen and air) and shifts the response to a lower temperature.

These results confirm earlier observations [10,11] that suggest that the presence of oxidative centers from processing or subsequent in-service aging or both promotes an auto-oxidative property. These cause shifts to lower temperatures with increased degree of oxidation. Subsequent research [6,7,8] has shown that the onset temperature, T_{on} of the oxidative exotherm, determined in air both in polypropylene and low density polyethylene, is a sensitive measure of the degree of oxidation introduced during simulated service-life histories.

Before considering the consequences of shifts in T_{on} values for polypropylene in particular, it is as well perhaps to compare these findings with those from the established oxygen-induction time or OIT method [ASTM Test Method for Oxidative-Induction Time of Polyolefins by Differential Scanning Calorimetry (D3895)]. The OIT is the time for the exothermic reaction to occur in an isothermal DSC or DTA (or mass loss in TGA) experiment [12]. Temperatures are chosen to give convenient experimental times of one hour or so and for polypropylene are of the order of about 160°C in air [4]. Foster [13] has used the technique to distinguish between the thermal stabilities of polyethylene of different origins for example. Use of OIT to distinguish between different antioxidant efficiencies is sometimes questioned because these measurements are carried out under molten polymer conditions while these stabilizers are designed to work in the heterogeneous solid state.

Recent studies by Day et al. [14] confirm not only our earlier studies [4] that kinetic constants of postfusion oxidative degradation are dependent upon polypropylene history, but also that OIT values and T_{on} values both reduce if the polymer has experienced high levels of previous oxidative degradation. No attempt, however, has been made to interpret shifts in T_{on} with respect to the nature of the auto-oxidative mechanism or with other polymer properties. This paper will examine the nature of these shifts in more analytical depth than hitherto.

THE EFFECT OF THERMAL OXIDATIVE HISTORY ON T_{ON} VALUES OF POLYPROPYLENE

Previously, we have reported [6,7] the changes in physico-chemical behavior of a number of oriented polypropylene filaments and tapes after exposure to elevated temperature (90 to 130°C) in an air oven for periods up to the failure or embrittlement time while experiencing varying levels of applied stress. While each set of exposed samples was examined by thermal analysis (DSC and TGA), the influence of thermal oxidative exposure on T_{on} values was examined only with relation to each set of experiments and not across the series.

Table 1 summarizes the selected sample characteristics and 130°C exposure conditions for each set of experiments in Ref 6 and 7.

TABLE 1 -- Polypropylene characteristics and exposure details for samples described in Ref 6 and 7.

Polymer as Filaments (f) or Tapes (t); (8:1 draw ratio)[a]	MFI (g per 10 min, at 130°C)	Exposure Conditions		
		Temp, °C	Maximum time, days [b]	Stress,[c] %
Ref 6:				
PP1(P1); f	12	130	27	0,5,10,15
PP2(P2); t	3	130	10	0,10,15
Ref 7:				
PP3(P1); t	3.0	130	8	0,10, 15
PP4(P2); t	3.8	130	10	0,5,10,15
PP5(P3); t	7.5	130	10	0,5,10,12.5
PP6(P4); t	12.4	130	60	0,5,10,12.5
PP7(PC); t	3.6	130	10	0,5,10,12,5

[a] Codes in parentheses are those in Ref 6 and 7.
[b] Maximum times are determined by the efficiencies of antioxidants present.
[c] Stress values are expressed as percent of respective ultimate breaking stress at 130°C.

In the two sets of experiments, polymers PP2, PP3, and PP4 have similar melt flow indices and antioxidant formulations. PP7 tapes are almost the same as PP4 tapes except that they contain 2% w/w carbon black and were of commercial origin; tapes from polymers PP1 to PP6 were all experimentally produced to commercial tensile specifications. The T_{on} values were determined by extrapolation (Fig. 2) of the exotherms to the extended baseline; DTA (or heat flux DSC) exotherms for all polymers in Table 1 were obtained using a Stanton Redcroft 671B instrument as previously described [4]. Similarly derived values were obtained by extrapolation of the respective DTG response (Fig. 3) for polymers PP3-PP7 using a Stanton Redcroft TG760 thermogravimetric analyser [4]. The T_{on} values derived from DTA/DSC measurements are collated (Table 2) (note that DSC values in Ref 7 were referred to as DTA - derived values; in essence the DTA technique used is heat flux DSC as described in Ref 6).

Similarly, TGA/DTG-derived values of T_{on} from oven-aged polypropylene tapes in Ref 7 and *15* are presented (Table 3).

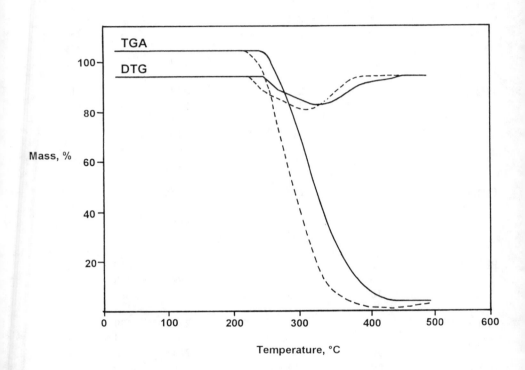

Fig. 3 -- <u>TGA/DTG responses in air at 10K min^{-1} for unaged (——) and zero stress, 10 day oven-aged (┈) PP4 tapes and 10% stress.</u>

The Effect of Thermal Exposure in Air on T_{on} values

Generally the results (Table 2) when plotted for zero stress conditions (Fig. 4) show that DSC-derived values of T_{on} decrease with increasing time of exposure and

TABLE 2 --   Collated DSC values (air, 10 K min^{-1}) of T_{on} from polypropylene filaments and tapes exposed in air at 130°C

Polymer Filament or Tape	Exposure Time, Days	T_{on} °C at Stress Levels				
		0	5	10	12.5	15.0%
PP1; f	0	241	-	-	-	241
	5	220	-	-	-	234
	10	210	-	-	-	231
	20	204	-	-	-	229
	27	a	-	237	-	239
PP2; t	0	214	-	214	-	214
	0.125	215	-	220	-	219
	1	217	-	221	-	212
	5	212	-	219	-	209
	7	-	-	-	-	-
	10	205	-	215	-	a
PP4; t	0	217	217	217	-	217
	3	-	-	-	-	196
	6	-	-	199	-	a
	10	204	202	a	-	
PP5; t	0	219	219	219	219	-
	6	-	-	199	208	-
	10	210.5	210	a	a	a
PP6; t	0	239	239	239	239	-
	10	-	-	-	234	-
	30	-	-	224	a	-
	60	214	227	a	-	-

a Denotes embrittlement or failure of tape before stated exposure time.

initially large reductions occur over relatively short time periods with respect to embrittlement or maximum exposure times or both. Furthermore, the application of stress can significantly reduce the magnitude of these T_{on} shifts especially in the case of PP1 filament and PP6 tapes which contain the more efficient antioxidant formulations [6]. In addition, the relative efficiencies of these stabilizer systems correspond to the respectively high T_{on} values before exposure (241 and 239°C, respectively). The effect of absence of stabilizers has been noted previously [7, 15] where the unstabilized reactor-grade parent of PP4 has a much lower T_{on} value of 196°C.

A similar set of conclusions may be drawn from the TGA/DTG-derived data (Table 3) although some unexpectedly high individual T_{on} values are seen. This may be because of the heterogeneity of the oxidation process noted previously by ourselves [15] and other workers [16]. This can present a problem when only small amounts of total sample are available for analysis as is the case here (for example, limited numbers of single filaments and tapes); this sampling may not be representative of the average degree of oxidation present in the polypropylene specimens. The case of PP5 is interesting in that all short-term exposed tapes show T_{on} rises from 216°C to values as high as 235°C. In absence of further data, this effect is difficult to explain unless short-term exposure is also promoting some kind of oxidative or even auto-oxidative termination reactions.

Figures 5 and 6 plot the results (Table 3) for all samples having the same antioxidant system in the absence (PP3, PP4, PP5) and presence (PP7) of carbon black, respectively. In common with the trend for DSC-derived T_{on} values (Fig. 4), and in spite of considerable scatter of data points, similarly decreasing trends are seen with increasing exposure time (and hence degree of oxidation). Simple curve-fitting suggests that a power law relationship (shown as the solid curves in Figs. 4, 5, and 6) may describe the trend although correlation coefficients are low (Table 5, see later). However, notwithstanding these reservations, there is an element of consistency in the data set trends.

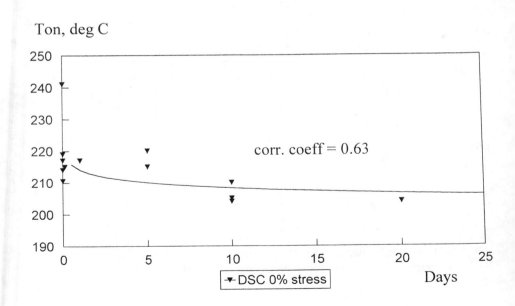

Fig. 4 -- DSC-derived T_{on} values versus exposure time.

TABLE 3 -- Collated TGA/DTG-derived values (air, 10 K min^{-1}) of T_{on} from polypropylene tapes exposed to 130° in air.

Polymer/ Tape	Exposure Time, Days	T_{on}, °C at Stress Levels				
		0	5	10	12.5	15
PP3	0	223				
	1	218	219	218.5	-	221.5
	4	212	209.5	219	-	a
	8	213.5	219	220	-	-
PP4	0	233				
	0.125	218	-	218	-	209
	1	220	216	226	-	215
	3	213	214	214.5	-	214
	6	210	210	215	-	a
	10	211	213	a	a	-
PP5	0	216				
	0.125	-	235	235.5	-	-
	1	235	231	224.5	213	-
	3	217	222	216	234.5	-
	6	213.5	215.5	211.5	221.5	-
	10	221	216	a	a	-
PP6	0	247				
	0.125	233	232	248	-	-
	1	224	245	242	242.5	-
	15	243.5	232.5	247	229(7)^b	-
	30	231	242	232.5	243.5(10)^b	-
	45	224.5	236	a	a	-
	60	228	231	-	-	-
PP7	0	230.5				
	0.125	-	232.5	234	-	-
	1	235	232	231	-	-
	3	206.5	220	219	-	132
	6	225.5	216	215	-	a
	10	216	214	214	-	-

a Denotes embrittlement or failure of tapes before stated time.
b Figures in brackets refer to reduced times of exposure in days.

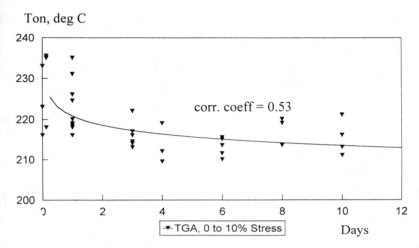

Fig. 5 -- <u>TGA-derived T_{on} values versus exposure time for PP3, PP4, and PP5.</u>

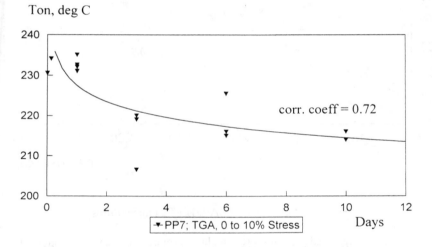

Fig. 6 -- <u>TGA-derived T_{on} values versus exposure time for PP7.</u>

Oxidation and Auto-oxidation Mechanisms

It is worth posing the question whether the DSC and TGA/DTG-derived onset temperatures of oxidation relate to the same thermal transition and hence describe the same oxidation processes. Some of the data above (Tables 2 and 3) refer to the same exposed samples for PP4, PP5, and PP6 polypropylene tapes. In addition, similar data exists for exposed tapes of the unstabilized analogue of PP4, which because of their relative oxidative sensitivity, were exposed for up to 21 days in air at only 90°C after which embrittlement occurred [7]. A plot of DSC - versus TGA/DTG-derived T_{on} values for all these identically exposed samples (Fig. 7), is approximately linear and indicates that both values do represent the same thermal transition although the shift in the best-fit line up the ordinate suggests that

$$T_{on}(TGA/DTG) > T_{on}(DSC)$$

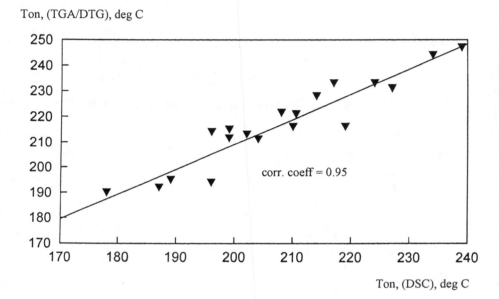

Fig 7 -- <u>DSC - versus TGA/DTG T_{on} temperatures</u>

We have suggested previously [6] and have produced evidence in support of competing oxidation reactions which have nonchain scission and chain scission characteristics. Furthermore, the former products are more resistant to oxidation and auto-oxidation than the latter. Based on our previous proposals and this analysis, a simplified view of this overall mechanism is presented (Fig. 8) where oxidative chain scission (r_s) competes with oxidative nonchain scission (r_{ns}); this competition is temperature - dependent as shown. Products of the latter subsequently undergo oxidative or auto-oxidative chain scission (r_s') at temperatures close to or above 250°C.

The DSC experiments show a multipeaked exotherm (Fig. 2) which suggests a multistage oxidation reflecting the various stages in Fig.8. TGA experiments, however, show one major volatilization DTG peak with a lower temperature shoulder (Fig. 3). A comparison of DSC and TGA/DTG principal maxima for a series of polypropylene samples (Table 3) are listed in Table 4; lower temperature exothermic shoulders or minor peaks are also indicated (Figs. 2 and 3).

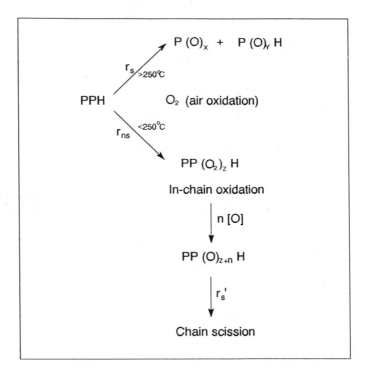

Fig. 8 -- <u>Competing oxidation and auto-oxidation reactions.</u>

Given that different instruments were used but that similar conditions (5mg sample mass, 10K min^{-1} heating rates in static air) were used and that the shape of DSC exotherm changes during oven exposure, the respective DSC maxima are less than TGA/DTG-derived maximum temperatures. They do, however, fall between the onset and maximum temperature of each respective major DTG peak and so may be associated with the principal volatilization stage occurring above 250°C with a maximum rate in the range 280 to 335°C. The consistently and slightly lower DSC T_{on} values with shoulder values close to or below 250°C could indicate that this transition represents the beginning of the nonchain scission oxidation reaction, r_{ns} (Fig.8) and at slightly higher temperatures, commensurate with the first TGA/DTG onset temperature, chain scission ensues (Fig. 8). Interestingly, the DSC and TGA/DTG shoulder temperatures are very similar (Table 4), and both probably define this first volatilization stage which is a consequence of oxidative and auto-oxidative chain scission of oxidation products of reaction r_{ns} (Fig. 2). It is proposed that this latter reaction is described by r_s' (Fig. 8). The major volatilization stage, however, occurs as a consequence of reaction r_s at temperatures above 250°C, and as measured by DSC or TGA/DTG, these are not as obviously influenced by polymer origin or oxidative history as are T_{on} values. It would appear, therefore, that it is the lower temperature region of the oxidative DSC exotherm and hence the nonchain scission, r_{ns}, and immediately subsequent chain scission oxidative reactions, r_s', that are most sensitive to polymer history and that the trends recorded previously and discussed above reflect this. This proposed mechanism supports previously published data regarding the effects of applied load and presence of antioxidant on thermal analytical behaviour [6]. This previous work suggested that oxidative chain scission occurred in competition with and also as a consequence of nonchain scission oxidation and that different antioxidants favoured preferential stabilisation of the various routes. Since applied load favours chain scission, reactions, differences in stress-sensitivity to oxidation were attributed to the relative abilities of antioxidants present to impede chain or nonchain scission oxidative routes.

TABLE 4 --   Comparison of temperatures of maximum exothermic intensity from DSC and TGA/DTG for selected polypropylene tapes exposed at 130°C in air [7].

Polymer/ Tape	Exposure Condition, Stress %/Days	Temperature of Maximum of Oxidative Exotherm or Mass Loss			
		DSC[a]		TGA/DTG[a]	
PP4			266	251[s],	322.5
	0/10		273	229[s],	335
	5/10	222[s],	284	231[s],	308.5
	10/6		262	229[s],	314
	15/3		273	229[s,]	282
PP5			295	251[s],	309
	0/10	237[s],	280	232[s],	322
	5/10	239[s],	291	231[s],	315
	10/6		266	231[s],	297.5
	12.5/6	222[s],	276	248[s],	306
PP6		251[s],	267/295	246[s],	302.5
	0/60	236[s],	277	248[s],	317.5
	5/60		273	239[s],	294.5
	10/30	234[s],	276	246[s],	289.5
	12.5/10		264	258[s],	323

a - s denotes shoulder or subpeak if distinctly identifiable.

THE EFFECT OF CARBON BLACK

More recent studies [17] of the thermal aging behavior of polypropylene tapes containing a range of carbon blacks confirm the above observations and especially that for PP7 (Fig. 6) which has the highest power law correlation coefficient. A selection of carbon blacks having different properties were introduced at 2.5% (w/w) concentration into the tapes (Table 5); those presented here comprise two smaller particle diameter carbon blacks with low (CBS1) and high (CBS2) volatile contents and two high (CBH1 and CBH2) particle-sized blacks with low volatile content. The volatile content is a measure of the degree of chemical oxidative modification and hence potential reactivity [17].

TABLE 5 -- <u>Properties of carbon blacks and tapes, containing 2.5% (w/w).</u>

Code	Particle	Volatile,	Tape Properties		
	Size, nm	%	Tenacity Ntex^{-1}	Breaking Strain, %	T_{on}(DSC)
PP	-	-	0.53	14	240
CBS1	16	1.5	0.37	19.3	255.5
CBS2	16	9.5	0.42	14.8	255.8
CBH1	60	1.0	0.52	18.6	246.9
CBH2	50	1.0	0.45	15	249.8

The unpigmented and pigmented tapes contained a different polypropylene to those used in the previous studies in terms of manufacturer, but it was similar with regard to MFI (1.8g/10min, 2.16 kg at 130°C) and antioxidant type. Tapes were produced as described previously [7,17] and had acceptable tensile and thermal properties (Table 5). The presence of carbon black is seen to increase the DSC-derived T_{on} temperatures of the polypropylene present, and this was noted for PP7 tapes (Table 3). In these experiments, DSC experiments were undertaken on a Polymer Laboratories instrument using 1.5-mg sample sizes under flowing air (10 mL min^{-1}) and heated at 10K min^{-1}. Oven aging at 130°C in air under zero stress promotes tape embrittlement following an induction period of about 20 days, and T_{on} values are plotted (Fig. 9). As with the former trends (Figs. 4, 5, and 6) so here, power law curve fitting was undertaken yielding correlation coefficients of 0.90 or better. Comparison of the reduction in T_{on} trends with respective losses in tensile properties, [17,18] demonstrates that again, the onset temperature of auto-oxidation is a good indicator of the degree of polymer oxidation and hence damage. Carbon blacks having low volatile contents change the thermal oxidative behavior of polypropylene less than one with a high volatile content. The dependence of particle size on thermal stability is not too clear and is currently under investigation.

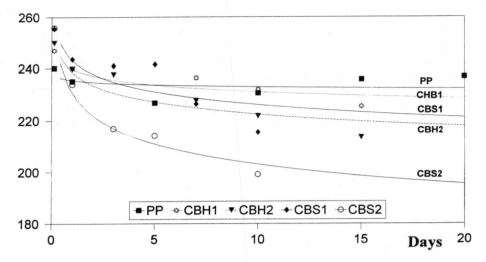

Fig. 9 -- <u>DSC-derived T_{on} values versus exposure time for carbon black-ontaining tapes</u>

TABLE 6 -- <u>Power law parameters.</u>

Sample/Techniques	a	b	Corr. Coefficient
PP1,2,4,5/DSC (Fig. 4)	5.37	-0.012	0.63
PP3-5/TGA (Fig. 5)	5.40	-0.015	0.52
PP7/TGA (Fig. 6)	5.43	-0.026	0.72
PP	5.46	-0.005	<0.1
CBS1	5.49	-0.031	0.95
CBS2	5.44	-0.055	0.90
CBH1	5.47	-0.016	0.96
CBH2 (Fig. 9)	5.47	-0.030	0.98

SIGNIFICANCE OF A POWER LAW RELATIONSHIP

The power law, $y = e^a x^b$ used to define the curve fits (Figs. 4, 5, and 9) from DTA/DSC and TGA/DTG measurements enables the tabulated fit parameters to be defined (Table 6).

While the correlation coefficients (and hence significances of fits) are low in the absence of carbon black, its presence increases the confidence of the power law relationships. In our earlier work [7], it was noted that changes in tape physical and tensile properties with time of exposure were more reliable in terms of experimental error and produced smoother trends. Similar observations exist with the more recent studies of carbon black filled tapes [17] as corroborated by the improved correlations (Table 6). Again in these studies, the less evident trend in the unfilled standard PP tape (Fig. 9) has prevented its significance being identified; hence, the absence of a correlation coefficient (Table 6). In spite of the variability of the various plots, a values are remarkably constant and perhaps reflect the fundamental polypropylene behavior. The b factor value vary and may be more indicative of the sensitivity of each tape to oxidation. The magnitude of b in the case of CBH1 and CBS2 reflects the lower and upper limits, respectively, of the sensitivity effect that the presence of carbon blacks has on the thermal oxidation of polypropylene with unfilled tapes having the lowest values of all ($b \leq -0.015$). CBS2 tapes also show the greatest loss in tensile properties during thermal aging [17].

CONCLUSIONS

The evidence and analysis of both previously reported and current research into the characterization of oxidative degradation in oven-exposed orientated polypropylene filaments and tapes shows that shifts in the thermoanalytically-derived onset temperatures of oxidation are sensitive indicators of extent of oxidation. Shifts in T_{on} values to lower values are observed well before significant changes in physical and chemical properties of the polymer occur. The origin of the DSC-derived oxidative transition is associated with nonscission chain oxidation; that associated with the slightly higher TGA/DTG-derived T_{on} values is a consequence of consecutive and simultaneous oxidative chain scission reactions (Fig. 8). Both T_{on} values for a given polypropylene filament or tape decrease asymptotically with time of oxidative exposure towards a lower, constant value by which stage significant changes in physical properties (for example, breaking strain and stress) have occurred.

While the presence of antioxidants and additive pigments, such as carbon black, influence the initial values of T_{on} (usually by increasing them), subsequent reductions in value occur at rates dependent upon their stabilizing ability following oxidative exposure. In all polypropylene samples studied, the fall in T_{on} fits a power law dependence with time of oxidative exposure.

This asymptoting behavior (Figs. 4, 5, 6, and 9) may be described qualitatively by a simple steadystate model in which oxidation of the polymer gives rise to oxidatively and possibly auto-oxidatively sensitive products PP(0_2)H, formed by nonscission reactions in the first instance (Fig. 8). The concentration of these determines the magnitude of the shift of T_{on} to lower temperatures. Prolonged polymer oxidation promotes both oxidative formation and auto-oxidative destruction of these initial products thereby giving rise to a steadystate population and hence a reduced, constant value of T_{on} at longer exposure times.

The above results and analyses are corroborated by our current research which has extended the work to low density [8] and linear low density polyethylene films, and in which the early stages of photodegradation have also been monitored as reductions in respective T_{on} values. Relative rates of reduction reflect relative stabilizing efficiencies of photostabilizers present in times much shorter than those required to promote mechanical failure. Therefore, it may be more generally anticipated that measurements of T_{on} of polyolefin filaments, tapes, and film during service may be useful in determining the extent of degradation present and, more importantly, in estimating the remaining respective useful lifetimes.

REFERENCES

[1]   Allen, N. S. and Edge, M., Fundamentals of Polymer Degradation and Stabilisation, Elsevier Applied Science, London and New York, 1992.

[2]   Gugumus, F., "Mechanisms of Photo-oxidation of Polyolefin," Die Angewandte Makromolekulare Chemie, Vol. 176/177, 1990, pp. 27-42.

[3]   Gugumus, F., "Thermo-oxidative Degradation of Polyolefins in the Solid State, Pts 1 - 3," Polymer Degradation and Stability, Vol. 52, 1996, pp. 119-170.

[4]   Horrocks, A. R. and D'Souza, J. A., "Physicochemical Changes in Stabilised, Orientated Polypropylene Films during the Initial Stages of Thermal Oxidation," Journal of Applied Polymer Science, Vol. 42, 1991, pp. 243-261.

[5]   Oswald, H. J. and Turi, E., "The Deterioration of Polypropylene by Oxidative Degradation," Polymer Engineering Science, Vol. 5, 1965, p. 152.

[6]   Horrocks, A. R., Valinejad, K., and Crighton, J. S., "Demonstration of the Possible Competing Effects of Oxidation and Chain Scission in Orientated and Stressed Polypropylene," Journal of Applied Polymer Science, Vol. 54, 1994, pp. 593-600.

[7]   Horrocks, A. R. and D'Souza, J. A., "The Effects of Stress, Environment and Polymer Variables on the Durabilities of Orientated Polypropylene Tapes," Polymer Degradation and Stability, Vol. 46, 1994, pp. 181-194.

[8]   Liu, M., Horrocks, A. R., and Hall, M. E., "Correlation of Physicochemical

Changes in UV-exposed Low Density Polyethylene Films Containing Various UV Stabilisers," Polymer Degradation and Stability, Vol. 49, 1995, pp. 151-161.

[9]   Bolland, J. A. and Gee, G., Transactions of the Faraday Society, Vol. 42, 1946, pp. 236-260.

[10]   Althouse, V. E., "Inhibition of Cross-linking and Oxidation by Some Common Antioxidants in Irradiated Polyethylene," American Chemical Society; Division of Polymer Chemistry, Vol. 4, 1963, pp. 256.

[11]   Hampson, F. W. and Manley, T. R., "A Thermoanalytical Comparison between Raw and Screw-extruded Polypropylene," Polymer, Vol. 17, 1976, pp. 723-726.

[12]   Reich, L. and Stivala, S. S., Elements of Polymer Degradation, McGraw-Hill, New York, 1971, p. 71.

[13]   Foster, G. N., "The Thermal Oxidative Stability of Polyethylene," in Proceedings of 7th International conference, Advances in Stabilisation and Controlled Degradation of Polymers, Technomic, Lancaster, Pa., USA, 1986, p. 9.

[14]   Day, M., Cooney, J. D., Klein, C., and Fox, J. L., "Use of Thermal Analysis to Study Thermal Processing Effects on Polypropylene," Journal of Thermal Analysis, Vol. 41, 1994, pp. 225-237.

[15]   D'Souza, J. A., "The Durability of Polypropylene Tapes for Use in Geotextiles," Ph.D. thesis, Council for National Academic Awards, Bolton Institute, UK, 1989.

[16]   Celina, M. and George, G. A., "A Heterogeneous Model for the Thermal Oxidation of Solid Polypropylene from Chemi-luminescence Analysis," Polymer Degradation and Stability, Vol. 40, 1993, pp. 223-336.

[17]   Mwila, J., "The Effect of Carbon Black on the Durability of Polypropylene Tapes for Geotextile Applications," Ph.D. thesis, University of Manchester, Bolton Institute, 1995.

[18]   Horrocks, A. R., Miraftab, M., and Mwila, J., "The Effect of Carbon Black on the Physical Properties of Polypropylene Geotextiles Tapes," to be published in the Proceedings of First European Geosynthetics Conference and Exhibition, Balkema, Rotterdam, 1996.

Yick, G. Hsuan[1] and Z. Guan[2]

EVALUATION OF THE OXIDATION BEHAVIOR OF POLYETHYLENE GEOMEMBRANES USING OXIDATIVE INDUCTION TIME TESTS

REFERENCE: Hsuan, Y. G. and Guan, Z., **"Evaluation of the Oxidation Behavior of Polyethylene Geomembranes Using Oxidative Induction Time Tests,"** Oxidative Behavior of Materials by Thermal Analytical Techniques, ASTM STP 1326, A. T. Riga and G. H. Patterson, Eds., American Society for Testing and Materials, 1997.

ABSTRACT: The oxidation of five different high density polyethylene (HDPE) geomembranes was produced in a forced air oven at a temperature of 115°C. The antioxidant packages of three of the geomembranes contained hindered phenols and phosphites types of antioxidants. The antioxidant packages of the other two geomembranes consisted of either thiosynergists or hindered amines together with hindered phenols and phosphites. The consumption of antioxidants after given time periods of oven incubation was monitored by two oxidative induction time (OIT) tests: standard OIT and high pressure OIT together with tensile tests. The results indicated that both OIT tests can effectively track the consumption of hindered phenol and phosphite types of antioxidants. However, for thiosynergists and hindered amines, the high pressure OIT is the appropriate test. The study clearly demonstrated that mechanical properties do not decrease until all (or most) of the OIT value has been depleted. Thus, the reduction of OIT values is the precursor of material degradation.

KEYWORDS: oxidative induction time (OIT), differential scanning calorimetry, polyethylene, geomembrane, oxidation

---

[1]Assistant Professor, Civil and Architectural Engineering, Drexel University, Philadelphia, PA 19104

[2] Graduate Student, Civil and Architectural Engineering, Drexel University, Philadelphia, PA 19104

76

Geomembranes are defined by ASTM D 4439 "Terminology for Geosynthetics" as "very low permeability synthetic membrane liners or barriers used with any geotechnical engineering related material so as to control fluid migration in a man-made project, structure, or system". Geomembranes are mainly used in environmental related applications which involve liners and covers of solid waste landfills, reservoir and canal liners, and heap leach pad liners. In most of these applications, longevity is an issue. Service life times of 30 to 100 years (and longer) are usually required. Thus, it is critical to develop an accelerated incubation protocol and subsequent test method(s) that can be used to assess long term performance.

Polyethylene is the most widely used polymer for the manufacturing of geomembranes. Both medium density and linear low density resins are used depending on the specific application. Since geomembranes are often covered by another material, such as soil, photo-degradation is not usually a concern. However, oxidation is clearly an issue regarding long-term durability. For protection against oxidation during the service period, antioxidants are added into the formulation of the geomembrane. The antioxidants extend the lifetime of the polymer by deactivating free radical species and decomposing the hydroperoxide to a stable alcohol [1]. However, there are many types of antioxidants. They obviously perform differently. The type of antioxidant used is certainly important with respect to the long term performance of the particular geomembrane. In addition to the types of antioxidants, the amount is also an important factor. The typical amount of antioxidants present in a geomembrane formulation is approximately 1.0 %.

Due to the many different antioxidant packages, a proper laboratory incubation and subsequent test method(s) to evaluate the effectiveness of the antioxidant is critical in estimating the longevity of the geomembrane. One of the most widely used laboratory incubation methods is oven aging. Samples of the manufactured geomembrane(s), are incubated in forced air ovens at an elevated temperatures for sufficient time so as to deplete the antioxidants and start the oxidative degradation process. Consequently, the physical/mechanical properties of the geomembrane begin to be degraded. The temperature of the ovens must be high enough to accelerate the aging process, but not so high as to exceed the melting point of the base polymer. Temperatures as high as 110°C and 120°C have been utilized in evaluating HDPE geomembranes by researchers [2, 3, and 4].

Incubated samples are periodically removed from the ovens and are then evaluated by selected tests which typically are mechanical tests, e.g., tensile, impact and bending tests. The time to reach 50% reduction in a mechanical property or the time to reach embrittlement is used to compare the performance of different antioxidant packages. However, the change of mechanical properties usually occurs abruptly. Furthermore, the times for change in mechanical properties can be very long. Most importantly, the test data does not provide information regarding the depletion of the antioxidants. Thus, an alternative test that can monitor the depletion of antioxidants is desired.

The oxidative induction time (OIT) test is a straight forward method to assess the amount of antioxidant present in the polymer. For the same antioxidant package (type and quantity), the OIT is proportional to the quantity of the antioxidant. Thomas and Ancelet [2] demonstrated that the high pressure oxidative induction time (OIT) decreased gradually as the oven incubation duration increased for twelve different antioxidant packages. The advantage of using the OIT test is that the antioxidant depletion rate of a particular geomembrane formulation can be established before the antioxidants are completely consumed. Hence, the total testing time can be significantly shortened. Therefore, a quality control (QC) and/or quality assurance (QA) test protocol can be developed by defining a specific OIT value for the newly manufactured geomembrane and then again after it is incubated in an oven for at a specified temperature and duration.

In this paper, five different high density polyethylene (HDPE) geomembranes were evaluated by two different OIT test methods: Standard OIT and High Pressure OIT. The applicability of these tests with respect to different antioxidant packages is discussed. In addition, the OIT depletion rates of the five geomembranes at oven temperature of 115°C are presented. A comparison is also made between the OIT depletion rates and the changes in mechanical properties.

TEST MATERIALS

Five different commercially available high density polyethylene (HDPE) geomembranes were included in this study. They are designated as Geomembranes A to E. All geomembranes were of the same thickness, i.e., 1.5 mm. The term "HDPE" requires further clarification. HDPE geomembranes are actually manufactured using virgin resin with a density between 0.932 and 0.940 g/ml. The additional 2 to 3 % of carbon black, however, increases the formulated density of the geomembrane to above 0.941 g/ml which is categorized as the high density polyethylene per ASTM D883. Thus, the term "HDPE" is used by the geomembrane industry although the base resin is actually MDPE.

These five geomembranes were produced by four manufacturers. They included both blown sheet and flat sheet extrusion processes. The resins utilized to manufacture the geomembranes were supplied by four resin producers. The physical properties of the geomembranes are listed in Table 1.

Regarding the antioxidant package in each HDPE geomembrane formulation; the quantity of each antioxidant component was not revealed by the manufacturers, however, the antioxidant types are believed to be those listed in Table 1.

Table 1 -- Physical properties and antioxidant packages in the HDPE geomembranes used in this study

Geomembrane	Density[1] (g/ml)	Melt Index[2] (g/10 min)	Antioxidant Types
A	0.9495	0.23	hindered phenols & phosphites
B	0.9493	0.15	hindered phenols & phosphites
C	0.9493	0.29	hindered phenols & phosphites
D	0.9492	0.15	hindered phenols, phosphites, & thiosynergists
E	0.9437	0.46	hindered phenols & hindered amines

Note: 1. This is the formulated density including carbon blacks and antioxidants per ASTM D 1505
         2. Per ASTM D 1238.

In selecting antioxidant types, a most important consideration is the effective temperature range. The antioxidant package should protect the geomembrane during both the high temperatures occurring in the extrusion process and the significant lower temperatures occurring during the service lifetime. For the four basic antioxidant types, the approximately effective temperature ranges are given in Figure 1.

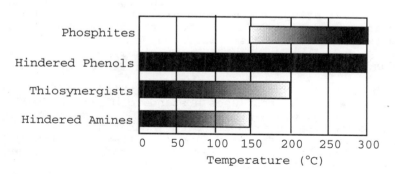

Fig.1 -- Effective temperature ranges of different antioxidant types. [5]

Figure 1 shows that phosphites have an effective temperature range above 150°C. They are considered to be process stabilizers. Contrary, thiosynergists and hindered amines have an effective temperature below 200°C and 150°C, respectively. They are used in a formulation to accommodate the low temperature service protection. For hindered phenols, they have a broad effective temperature range, from service temperatures through process temperatures. Thus, they can be used to protect the polymer in both situations.

For antioxidants that have a low effective temperature range, they will be degraded or volatilized at temperatures above their maximum range. This is the compelling reason for developing the high pressure OIT test since it is conducted at a lower temperature than is the standard OIT test.

INCUBATION TEST PROCEDURES

The oven aging incubation procedure for this study was performed according to ASTM D5271. Coupons with dimensions of 100 mm wide and 180 mm long were taken from each geomembrane sample. Each coupon was freely hanging in a force air oven maintained at a temperature of 115°C.

Small specimens were taken from the incubating coupons at monthly intervals for OIT evaluation. Specimens for mechanical tests were taken from the incubating coupons at three months intervals.

OXIDATIVE INDUCTION TIME TESTS

There are two ASTM OIT test methods to determine the amount of antioxidants in polymers such as HDPE geomembranes. One is ASTM D 3895 which is the Standard OIT (Std-OIT) test. The test is performed using a standard differential scanning calorimetry cell. The other test is ASTM D 5885 which is the High Pressure OIT (HP-OIT) test. The test uses a high pressure DSC cell. The test conditions of the two OIT methods used for evaluating geomembranes are summarized in the following subsections. In addition, the differences between these two methods are also discussed.

Standard Oxidative Induction Time (Std-OIT) Test

The Std-OIT test was performed in this study according to ASTM D3895. The test uses a standard DSC cell which can sustain a 35 kPa gauge pressure. Five (5) mg geomembrane specimens were taken from the oven incubating coupons at locations that were at least 5 mm away from the edges. In the DSC cell, the specimens were heated from room temperature to 200°C at a heating rate of 20°C/min under a nitrogen atmosphere. The gas flow rate was maintained at 50 ml/min. When 200°C was reached, the cell was maintained in an isothermal condition for 5 minutes. The gas was then changed to oxygen with a gauge pressure of 35 kPa and 50 ml/min flow rate. The test was terminated after an exothermal peak was detected. The OIT time was defined as the length of time between the introduction of oxygen and the onset of the exothermal peak. The minimum OIT value that could be reasonably measured was approximately 0.5 minute.

The majority of incubated coupons were evaluated by a single OIT test. Duplicate tests were performed on the non-incubated coupons and on selected incubated coupons to verify the consistency of the test, particularly for coupons with low OIT values.

High Pressure Oxidative Induction Time (HP-OIT) Test

The HP-OIT test procedure was performed according to ASTM D5885, with a minor modification. The test utilized a DSC cell that can sustain a pressure of 3500 kPa. Five (5) mg geomembrane specimens were taken from the incubating coupons at locations that were at least 5 mm away from the edges. The specimens were heated from room temperature to 150°C at a heating rate of 20°C/min under a nitrogen atmosphere. The pressure of the cell in this nitrogen stage was maintained at 35 kPa gauge pressure. The gas flow rate was not monitored. (The ASTM standard requires the DSC cell to be saturated with oxygen at a pressure of 3500 kPa before heating is started). When 150°C temperature was reached, the cell was maintained in an isothermal condition for 5 minutes. The gas was then changed from nitrogen to oxygen. The oxygen pressure in the cell was gradually increased to 3500 kPa within 1 minute. The test was terminated after an exothermal peak was detected. The minimum OIT value that could be reasonably measured was approximately 30 minutes.

As with the Std-OIT test, the majority of incubated coupons were evaluated by a single OIT test. Duplicate tests were performed in the non-incubated coupons and on selected incubated coupons to verify the consistency of the test, particularly for coupons with low OIT values.

The 150°C isothermal temperature is specified because HDPE geomembranes reach complete melting at approximately 135°C. The thermal oxidation of the specimen should proceed in a fully molten stage in order to eliminate any residual effects from the manufacturing processes. In addition, this is the highest temperature one should consider to avoid degradation of hindered amine antioxidants, recall Figure 1. Regarding the cell pressure, Tikuisis, et. al., [6] found that at an isothermal temperature of 150°C and pressure greater than 3500 kPa resulted in little change in HP-OIT values. Thus a pressure of 3500 kPa is stipulated in the standard.

Commentary on the Different Oxidation Induction Time Tests

As described in the previous subsections, the major
differences between the standard and high pressure OIT tests
are the pressure and isothermal temperature. The differences
create somewhat of a dilemma insofar as the selection of a
preferred test method for measuring OIT. Table 2 summaries
the advantages and disadvantages of each method.

Table 2 -- Differences between the Std-OIT and
HP-OIT test methods

Test	Advantages	Disadvantages
Std-OIT	- Short testing time. (<200 min.) - Simply to operate.	- High temperature may bias the test results for certain types of antioxidants.
HP-OIT	- Able to distinguish different types of antioxidants - Lower temperature relates closer to service conditions	- Long testing time (>300 min.) - Special testing cell and set up are required

MECHANICAL TESTS

The tensile test that was used to evaluate the
mechanical properties of the oven incubated HDPE geomembranes
was performed according to ASTM D638, Type V. Three dumbbell
shaped specimens were taken from each oven incubated coupon
on three month intervals. All specimens were taken
perpendicular to the machine direction.
The tensile test was performed at a strain rate of 0.2
mm/sec. The average value (3 tests) of the Young's modulus,
break stress and the break strain were obtained and were used
in the data analysis. The yield stress and yield strain
values were recorded but they were not analyzed due to their
lack of sensitivity to the thermal oxidation degradation.
There was no expected variation in yield properties, nor was
any variation measured.

Results of Std-OIT and HP-OIT Tests

The oxidative induction time of the non-incubated
specimen and the incubated specimens were monitored by both
Std-OIT and HP-OIT tests.   The results of each of the tests
are discussed separately.

Std-OIT Test Results

Table 3 lists the Std-OIT values of non-incubated geomembrane specimens. All Std-OIT values were very similar (they ranged from 122 to 156 min.), even though they contained different antioxidant packages.

Table 3 -- Comparing Std-OIT and HP-OIT values of unaged geomembranes.

Geomembrane	Std-OIT (min)	HP-OIT (min)	HP/Std
A	122	262	2.1
B	129	280	2.2
C	156	368	2.4
D	126	1125	9
E	144	1499	10

The Std-OIT values after 240 days oven incubation time are illustrated in Figure 2. Note that the results have been normalized so that the curves can be superimposed. The OIT values of all five geomembranes decreased exponentially over time. The reduction behavior was very similar in most geomembranes, except for Geomembrane E in which the OIT value decreased slightly faster than the others. The OIT values approached zero at 240 days in the majority of the geomembranes. Geomembrane D still retained a 10% OIT value at 240 days.

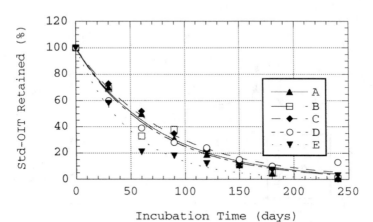

Fig.2 -- Changes in percent retained of Std-OIT values

Based on Std-OIT results, it seems that the different types of antioxidants did not have a significant effect on

either the initial OIT values or the rates of reduction of
OIT values. (It will be shown later that the geomembranes
did behave differently, particularly for Geomembranes D and
E. The correlation between OIT and tensile properties will
be discussed later in the paper). Thus, an alternative OIT
test method which can provide a better distinction between
the performance of various types of antioxidants is desired.
The HP-OIT test seems to be a possible candidates method.

<u>HP-OIT Test Results</u>

     Listed in Table 3 is the HP-OIT values of the non-
incubated geomembranes. The most noticeable difference
between these HP-OIT values are Geomembranes D and E which
showed much higher OIT values than Geomembranes A, B and C.
This difference was due to the additional thiosynergists and
hindered amines in Geomembranes D and E, respectively.
     Figure 3 shows the HP-OIT retained values plotted
against the monthly incubation time. The HP-OIT values
generally decreased slowly at early incubation time, then
became quicker as the amount of antioxidants diminished.
Geomembranes A, B and C exhibited similar reduction
behaviors. Their OIT values decreased to an undetectable
level at 240 days. Geomembranes D and E behaved quite
differently. Geomembrane D exhibited the greatest initial
reduction in OIT, a 70% drop, but after that the OIT value
remained almost constant throughout the next 180 days.
Geomembrane E showed the least OIT reduction of all. There
was still a 60% OIT retained in the material after 240 days
of incubation.

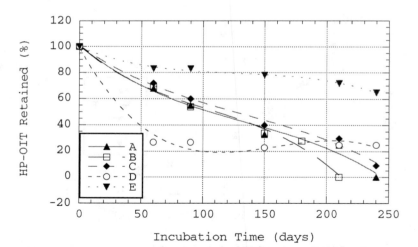

Fig.3 -- Changes in percent retained of HP-OIT values

Comparison Between Std-OIT and HP-OIT

In Table 3, the ratio of HP-OIT to Std-OIT values are listed. The HP-OIT values are always higher than the corresponding Std-OIT values due to the low isothermal test temperature. However, there was no single ratio factor for all geomembranes. Geomembranes A, B and C, which contained similar antioxidant packages, have similar ratio values. For Geomembranes D and E, a very high ratio is obtained. These high values are probably due to the additional low temperature antioxidants; thiosynergists and hindered amines. These two antioxidants performed effectively at the 150°C HP-OIT test temperature but to a much less extent at the 200°C Std-OIT test temperature.

The contrast between Std-OIT and HP-OIT reduction behaviors during oven incubation can be performed by presenting their OIT percent retained values on the same graph, as seen in Figures 4 to 8. Geomembranes A, B and C exhibited a similar depletion behavior by both tests. For Geomembrane D, a rapid reduction in the HP-OIT was detected within first 30 days of aging, and then the OIT value remained constant. Similarly, its Std-OIT value also leveled off at 150 days after a gradual decrease. The biggest difference between Std-OIT and HP-OIT was observed in Geomembrane E. As can be seen in Figure 8, the Std-OIT decreased rapidly and approached zero at 240 days. However, the HP-OIT still exhibited 60% retained OIT within the same incubation period.

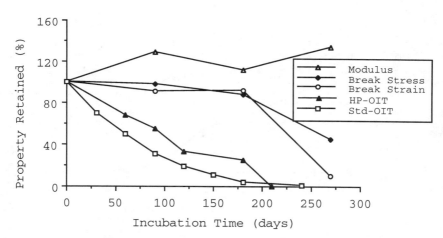

Fig. 4 -- Percent Retained of Properties for Geomembrane A

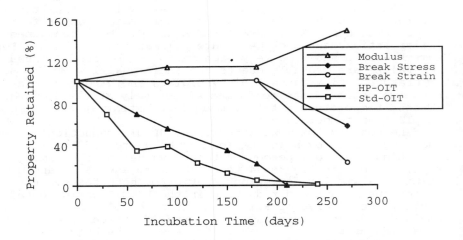

Fig. 5 -- Percent Retained of Properties for Geomembrane B

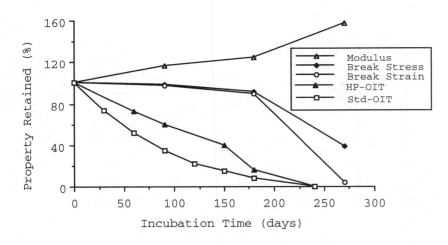

Fig. 6 -- Percent Retained of Properties for Geomembrane C

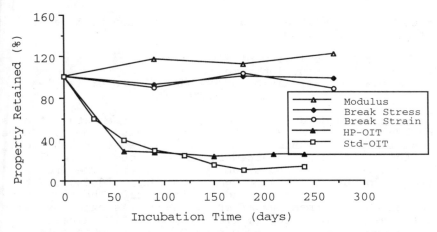

Fig. 7 -- Percent Retained of Properties for Geomembrane D

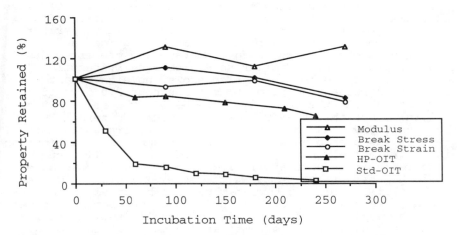

Fig. 8 -- Percent Retained of Properties for Geomembrane E

COMPARISON BETWEEN OIT AND TENSILE PROPERTIES

The retained tensile strength properties of each geomembrane was measured and superimposed onto the corresponding OIT values, see Figures 4 to 8. In assessing these graphs collectively, the most important finding is that modulus, break stress and break strain do not decrease until all (or most) of the OIT value has been depleted. Clearly, the existence of measurable OIT values is a precursor of any mechanical degradation which cannot occur until the antioxidants have been consumed. This type of multiple stage lifetime of geomembranes has been postulated by Hsuan and Koerner [7] as follows:

Stage A - Depletion of antioxidants
Stage B - Induction time
Stage C - degradation of mechanical properties leading
         to "halflife" of the material

Regarding details of Figures 4 to 8, the data reflects that Geomembranes A, B and C (which contained the same types of antioxidants) exhibited a similar degradation behavior. The tensile modulus remained relatively constant until 180 days and then increased slightly thereafter. The break stress and strain remained unchanged until 180 days, after which both properties dropped significantly. The break strain reduced to less than 20% after 270 days. As expected, within this time period of 180 to 270 days, both OIT values approached zero. For Geomembrane D, the three tensile parameters remained almost unchanged throughout 270 days of aging, in spite of a significant reduction in both OIT values. This observation suggests that the initial OIT loss in Geomembrane D might be due to the volatilization of certain types of antioxidant at the 115°C oven temperature. However, the remaining antioxidant still provided an effective protection to the geomembrane and preserved its tensile properties. Similar to Geomembrane D, tensile properties of Geomembrane E also showed no significant changes in mechanical properties throughout 270 days of incubation. At the same time, the HP-OIT was still 60% of original, but the Std-OIT had already reached zero. This contradictory behavior between tensile properties and Std-OIT results was probably caused by the consumption of antioxidant in the relatively high temperature of the Std-OIT test. Hindered amines which are included in Geomembrane E, have their highest effective range of temperature at 150°C. At 200°C, Std-OIT testing temperature, the hindered amines were rapidly degraded. Hence, the Std-OIT curve probably reflects only the depletion of the hindered phenolic antioxidants which were completely consumed at 240 days similar to Geomembranes A, B and C. Conversely, the HP-OIT response in Geomembrane E, closely substantiates the lack of change in mechanical property which is still above 60% after 270 days incubation.

SUMMARY AND CONCLUSIONS

The thermal oxidation of five different geomembranes was simulated using a forced air oven at a temperature of 115°C. The progress of the ongoing oxidation during 270 days of incubation was monitored using Std-OIT, HP-OIT and tensile tests. The test results can be summarized as follows:

Regarding the OIT testing:

1. Both Std-OIT and HP-OIT tests can effectively track the consumption of antioxidants in HDPE geomembranes.
2. For geomembranes that contain hindered phenol and phosphite antioxidants, both Std-OIT and HP-OIT are appropriate tests.
3. For geomembranes that contain thiosynergists or hindered amines, the Std-OIT test is not suitable due to its high isothermal testing temperature.
4. For geomembranes that contain thiosynergists or hindered amines, the HP-OIT test is recommended. The conditions that are suggested for the testing are at 150°C temperature and 3500 kPa pressure.

Regarding the antioxidant behavior:

1. The antioxidants in the formulation must be essentially be consumed before mechanical property degradation begins to occur.
2. The type of antioxidant plays a critical role in the anticipated lifetime of HDPE geomembranes.
3. The addition of thiosynergists or hindered amines appear to extend life with respect to hindered phenols and phosphites.

Thus the conclusion of this study is the recommendation for the use of HP-OIT when the antioxidant package in a HDPE geomembrane formulation is not known, e.g., for conformance testing or quality assurance work. Conversely, if the antioxidant package is known, e.g., for manufacturers, then either Std-OIT or HP-OIT can be used in quality control work.

From the perspective of the designer and user, this study has clearly demonstrated that essentially complete antioxidant consumption is required before mechanical property degradation begins. This accentuates the importance of antioxidants in the geomembrane formulation and their identification via the above recommended OIT measurements.

ACKNOWLEDGMENT

This research is supported by the National Science Foundation, via its Geomechanical, Geotechnical and Geo-Environmental Systems (G3S) program under Grant No. CMS-9312772.

REFERENCES

[1]   Grassie, N.and Scott,G.,<u>Polymer Degradation and Stabilization</u>, Published by Cambridge University Press, New York, USA (1985)

[2]   Thomas, R.W., and Ancelet, C.R., "The Effect of
      Temperature, Pressure and Oven aging on the High-
      Pressure Oxidation Induction Time of Different Types of
      Stabilizers", Geosynthetics'93 Conference Proceedings,
      Vancouver, British Columbia, Canada, Published by IFAI,
      St. Paul, MN, 1993, pp 915-924.

[3]   Yim, G., and Godin, M., "Long-Term Heat Aging
      Stabilization Study of Polyethylene and Its Relationship
      with Oxidative Induction Time (OIT)", Geosynthetics'93
      Conference Proceedings, Vancouver, British Columbia,
      Canada, Published by IFAI, St. Paul, MN, 1993, pp 803-
      815.

[4]   Gary, R.L., "Accelerated Testing Methods for Evaluating
      Polyolefins Stability", Geosynthetic Testing for Waste
      Containment Applications, ASTM STP 1081, Robert M.
      Koerner, editor, American Society for Testing and
      Materials, Philadelphia, 1990, pp 57-74.

[5]   Fay, J.J., and King, R.E., "Antioxidants for
      Geosynthetic Resins and Applications", Geosynthetics
      Resins, Formulations and Manufacturing, Edited by Hsuan,
      Y.G., and Koerner, R.M., GRI Conference Series,
      Published by IFAI, St. Paul, MN., 1994, pp 77-96.

[6]   Tikuisis, T., Lam, P., and Cossar, M., "High Pressure
      Oxidative Induction Time Analysis by Differential
      Scanning Calorimetry", MQC/MQA and CQC/CQA of
      Geosynthetics, Edited by Koerner, R.M., and Hsuan, Y.G.,
      GRI Conference Series, Published by IFAI, St. Paul, MN.,
      1993, pp 191-201.

[7]   Hsuan, G.Y. and Koerner, R.M., "Long Term Durability of
      HDPE Geomembranes, Part I - Depletion of Antioxidants"
      Geosynthetic Research Institute, Internal Report #16,
      December, 1995, pg. 37.

Alan T. Riga[1], Ricardo M. Collins[1], and Gerald H. Patterson[1]

Thermal Oxidative Behavior of Readily Available Reference Polymers

Reference: Riga, A. T., Collins, R. M., and Patterson, G. H., "**Thermal Oxidative Behavior of Readily Available Reference Polymers**," *Oxidative Behavior of Materials by Thermal Analytical Techniques, ASTM STP 1326*, A. T. Riga and G. H. Patterson, Eds. American Society for Testing and Materials, 1997.

Abstract: There is an ongoing need in the plastics and rubber industries for readily available polymer oxidation standards. Appropriate and well characterized polymers are needed as ASTM and industrial standards for a variety of applications, such as oxidative stability of engineering plastics, composites, coatings, and elastomers. Oxidative properties by thermoanalytical techniques, with an established polymer set, are of utmost importance in presenting reliable data and developing quality assurance protocols. Control charting thermal oxidative methods with quality polymers can lead to International Standards Organization (ISO) certification.

Thermal oxidative behavior of a commercial set of reference polymers has been evaluated by thermogravimetric analysis (TGA) differential thermal analysis (DTA), differential scanning calorimetry (DSC), and pressure DSC (PDSC). There is a good correlation between measured stability by PDSC in oxygen and DTA in air. Known polymer crystallinity properties have been used to calibrate test temperature and heats of fusion. The thermal oxidative properties of these readily available reference polymers are repeatable and reproducible.

Keywords: oxidation, oxidation induction time (OIT), oxidation temperature (OT), peak melt temperature, heat of fusion, reference polymers, scanning temperature technique, air-oxidation, differential scanning calorimetry (DSC), pressure differential scanning calorimetry (PDSC), polyolefins, oxidative stability, isothermal temperature, standard deviation, precision

There is an ongoing need in academic or industrial polymer research for standard test methods to determine the oxidative properties [1,2]. There is an equal need for readily available polymer references. ASTM Committees D20 on Plastics and E37 on Thermal Methods continuously require polymer standards to establish new test methods. There is an ASTM Committee D20 Test Method for Oxidative Induction Time of Polyolefins by Differential Scanning Calorimetry(D3895). There is also an ASTM Committee D2 Test

---

[1] Senior research chemist, contract engineer, and technology manager, respectively, The Lubrizol Corporation, 29400 Lakeland Blvd., Wickliffe, OH 44092

Method for Oxidation Induction Time of Lubricating Grease by Pressure Differential
Sanning Calorimetry (D5483). Polyethylene plastic pipe and fitting materials can be
characterized by a DSC oxidative stability method, ASTM Specification for Polyethylene
Plastics Pipe and Fittings Materials (D3350). DSC is also used in ASTM Test Methods
for Physical and Environmental Performance Properties of Insulations and Jackets for
Telecommunications Wire and Cable (D4565). ASTM Committee E37 has recently
completed an OIT test method and precision statement [3]. The commercial polymers
used in this study have been fully thermally characterized by TGA, DSC,
thermomechanical analysis (TMA), X-ray diffraction analysis (XRD), and Fourier
transform infrared analysis (FTIR) [4].

The focus of this research is to determine the oxidative properties of readily available
reference polymers by DSC and PDSC using ASTM Committee E37 standard test
methods. The OIT, as stated in the test method [3], "is a relative measure of oxidative
stability at the test temperature and is determined from data recorded during the
isothermal DSC test." The OIT values obtained are compared from one olefinic polymer
to another to obtain relative oxidative stability for the series of polymers studied. A new
DTA air-oxidation test was developed. A good correlation, r2=0.90, for 8 reference
polymers, has been established between the PDSC OIT data and a DTA oxidation onset
temperature, OOT, for the same polymer series.

Experimental Procedures
A TAI model 910 differential scanning calorimeter or pressure differential scanning
calorimeter was used in this study. The experimental conditions were as follows: sample
size 3.0 mg, oxygen flow rate 50 mL/min, heating rate of 40°C/min to an isothermal
temperature of 175°C, and a pressure of 3.5 MPa (500 psig) for PDSC and 195°C and 10
kPa pressure for DSC. Specimen capsules were aluminum pans that were clean, dry, and
flat. The polymer melt temperature (peak temperature, °C), heat of fusion (J/g), and OIT
(min) were measured in the DSC and PDSC tests.

A Seiko-Haake RTG, model 220, thermogravimetric analyzer-differential thermal
analyzer was used for the oxidation study in air. The experimental conditions were as
follows: sample size 10 mg, air flow rate 250 mL/min, heating rate of 10°C/min to
600°C, and platinum sample pans. The DTA polymer melt temperature (peak
temperature, °C), and oxidation onset temperature, OOT (as the exothermic extrapolated
onset temperature) were measured in the DTA test.

The materials used in this study were selected polymers, polyethylene, polypropylene,
and other olefinic polymers from the ResinKit Company[2], (see Table 1). The temperature
output of the DSC or PDSC was calibrated using test method E967 [11], except that a

[2] Resin Kit Co., P O Box 509, Woonsocket, RI 02895.

TABLE 1 - *Description of Reference Polymers*

Sample	Description
"D" HDPE	High Density Polyethylene, ASTM Reference D
#44 PP/Talc	Polypropylene with Talc, Polymer Resinkit #44
#27 PP	Polypropylene, Polymer Resinkit #27
#34 EVA	Ethylene Vinyl Acetate, Polymer Resinkit #34
#49 MDPE	Medium Density Polyethylene, Polymer Resinkit #49
#25 HDPE	High Density Polyethylene, Polymer Resinkit #25
#38 PP/FI Retrd	Flame Retardant Polypropylene, Polymer Resinkit #38
#24 LDPE	Low Density Polyethylene, Polymer Resinkit #24
#26 PP/co-poly	Polypropylene Co-polymer, Polymer Resinkit #26
#45 PP/CaCO₃	Polypropylene with Calcium Carbonate, Polymer Resinkit #45
#46 PP/Mica	Polypropylene with Mica, Polymer Resinkit #46
#22 Polybutylene	Polybutylene, Polymer Resinkit #22
#43 Polymethyl Pentene	Polymethyl Pentene, Polymer Resinkit #43

heating rate of 1°C/min. is used to approach the isothermal conditions of this test. Indium and tin calibration materials are used to bracket the isothermal; temperatures in the PDSC (175°C) and DSC (195°C) tests. The melting temperatures observed in the DSC or PDSC calibration are obtained from the extrapolated onset temperatures. The E37.01.10 ASTM polymer reference D used in the Interlaboratory study to establish precision was also used to calibrate the DSC, PDSC and DTA systems.

Results and Discussion
Polymer melt temperatures were used to verify the temperature scale under oxidative operational conditions. ASTM polyethylene reference D was a reliable and consistent polymer for the interlaboratory study. Representative PDSC curves for a "good" oxidatively stable polyolefin, reference D, are seen in Figs. 1 and 2. Typical PDSC curves for a "poor" oxidatively stable polyolefin, polybutylene, are seen in Figs. 3 and 4. OIT ranking is based on interlaboratory round robin test results and experimentally designed OIT experiments [1,2]. A poor OIT reference polymer is, considered in the context of this paper, a polymer with an OIT of <8 min, a fair OIT reference is 10 to 20 min and a good OIT reference is > 20 min, i.e., 27 to 41 min in this study, see Table 2.

TABLE 2 - *Oxidative Behavior of Reference Polymers*

Sample	Run	Toe, °C	Tp, °C	OIT, min	OIT Status
"D" HDPE	1	117.2	127.4	36.2	Good
	2	121.8	129.6	32.1	Good
#44 PP/Talc	1	158.0	167.4	40.4	Good
	2	157.1	168.1	41.4	Good

#27 PP	1	158.3	169.5	38.7	Good
	2	161.5	167.2	32.9	Good
#34 EVA	1	78.9	92.3	27.3	Good
	2	81.3	96.5	29.2	Good
#49 MDPE	1	120.8	139.3	28.9	Good
	2	121.9	136.4	39.1	Good
#25 HDPE	1	124.8	138.7	30.4	Good
	2	124.5	132.6	29.5	Good
#38 PP/FI Retrd	1	159.4	166.0	19.9	Fair
	2	157.6	168.0	19.3	Fair
#24 LDPE	1	96.0	104.8	17.2	Fair
	2	95.1	104.9	17.9	Fair
#26 PP/co-poly	1	146.3	163.7	11.2	Fair
	2	154.8	164.3	13.6	Fair
#45 PP/CaCO$_3$	1	159.1	168.2	7.2	Poor
	2	159.8	166.2	8.5	Poor
#46 PP/Mica	1	160.1	166.7	7.4	Poor
	2	155.5	165.6	7.5	Poor
#22 Polybutylene	1	107.5	126.2	6.9	Poor
	2	112.4	124.5	6.7	Poor
#43 Polymethylpentene	1	*	--	--	**

(Ramp 40°C/min from 25 to 195°C; Iso @ 195°C for 60 min; P = 3.5MPa)
* = Exothermic reaction; Al pan melted   ** = Not Recommended

Average sample weight used was 3.0 mg.
Toe = onset temperature for melting.
Tp = peak temperature for melting.
OIT = oxidation induction time where
          good = represents OIT > 20 min,
          fair = represents 10 < OIT > 20 min, and
          poor = represents OIT < 10 min.

The OIT by PDSC for ASTM reference D is 36.2 min (see Fig. 5). The OIT by PDSC for polybutylene, #22, is 6.9 min (Fig. 6). Twelve ResinKit polyolefins were examined by PDSC, and their OIT values are summarized in Fig. 7 and Table 2. Four of the twelve ResinKit polyolefins examined by PDSC were also evaluated by DSC at 195°C (Table 3).

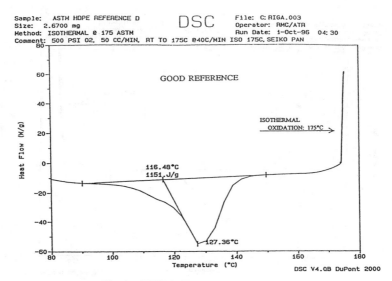

Figure 1: PDSC of ASTM HDPE Reference D-Overview

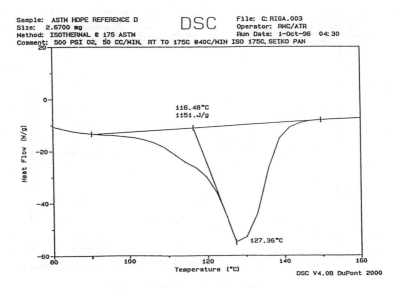

Figure 2: PDSC of polybutylene #22-Overview

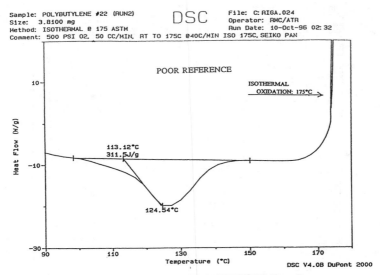

Figure 3: Melt profile of ASTM HDPE Reference D

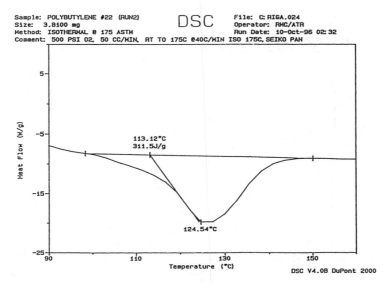

Figure 4: Melt profile of Polybutylene #22

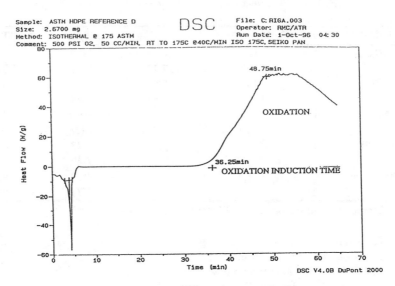

Figure 5: PDSC OIT of ASTM HDPE Reference D

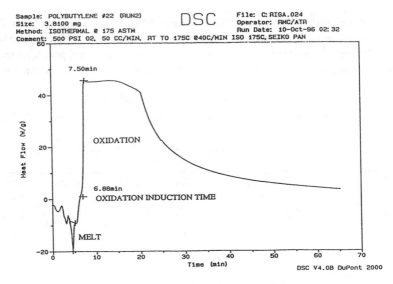

Figure 6: PDSC OIT of Polybutylene #22

TABLE 3 - *Oxidative Behavior of Reference Polymers by DSC*

Sample	Run #	Toe, °C	Tp, °C	OIT, min	OIT Status
"D" HDPE	1	118.8	128.4	22.8	Good
	2	119.3	137.1	31.1	Good
#27 PP	1	145.6	165.6	28.5	Good
	2	156.2	137.8	34.8	Good
#49 MDPE	1	122.5	136.0	10.8	Fair
	2	126.1	136.4	11.7	Fair
#24 LDPE	1	92.4	106.5	9.78	Poor
	2	98.3	109.9	7.47	Poor
#26 PP/co-poly	1	151.4	162.6	7.99	Poor
	2	151.1	162.6	7.77	Poor

(Ramp 40°C/min from 25 to 195°C; Iso @ 195°C for 60 min; 10kPa)

Toe = onset temperature for melting
Tp = peak temperature for melting
OIT = oxidation induction time where
     good = represents OIT > 20 min,
     fair = represents OIT < 20 min, and
     poor = represents OIT < 10 min
Average sample size used was 2.7 mg and average standard deviation was ±0.6 mg.

Five polyolefins had OIT values by PDSC greater than 20 min. They were considered oxidatively stable and good reference polymers: polypropylene, polypropylene with a talc filler, high density polyethylene, medium density polyethylene, and polyethylene vinyl acetate. Polymethylpentene was oxidatively unstable, since it melted and decomposed.

The melt properties of ASTM reference D by DSC and PDSC are summarized in Table 4. The peak temperature was 131°C, ±3.1% and the heat of fusion was 1040 J/g ±5.2%.

TABLE 4 - *Statistical Analysis of ASTM Reference D Melt Properties*

Run #	ΔH	Tp	Average ΔH	Std. Dev. ΔH	% Std. Dev. ΔH	Average Tp	Std. Dev. Tp	% Std. Dev. Tp
1	1150	127	1040	54	5.2	131	4.0	3.0
2	1050	130						
3	1060	128						
4	950	137						

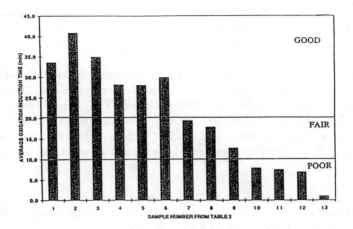

Figure 7: Oxidation induction time of reference polymers

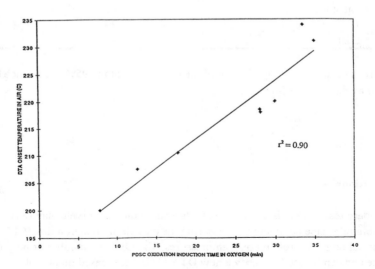

Figure 8: PDSC oxidation induction time versus DTA oxidation temperature

| 5 | 1070 | 127 |
| 6 | 970 | 134 |

ΔH is the heat of fusion in J/g

Tp is the peak melting temperature in °C

The peak temperature of reference D was used to calibrate the DTA air oxidation scanning temperature technique. The oxidation onset temperature, OOT, as defined by DTA in air correlated well with the OIT by PDSC at 175°C and 3.5 MPa oxygen (Figure 8). The eight polymer correlation coefficient (r^2) comparing OIT and OOT was *0.90*. This good correlation suggests that the reference D, high density polyethylene, as well as the other polyolefins, oxidatively degrade in the same manner under high pressure oxygen or low pressure air.

A comparison of OIT values for ASTM reference D based on the ASTM Committee E37 and D20 interlaboratory studies and the OIT values determined in this study are in good agreement (Table 5). Overall the OIT for reference D is 30 min ± 2.6 min.

TABLE 5 - *Oxidation Induction Time (min)* *

| Temperature | 175°C | 195°C | 200°C |
Pressure	3.5 MPa	10 kPa	10 kPa
Current study	33	27	--
E37	26	29	--
D20	--	--	31

* There is no statistical significance of the OIT values at the 95% confidence level: overall average OIT = 30 min ±2.6 min.

Conclusions

A standard test method has been used to determine the oxidative behavior of readily available reference polymers by measuring the oxidation induction time(OIT) at selected isothermal temperatures and oxygen pressures. The OIT of a number of polyolefins is precise and can be used to verify a new DTA (TGA) test based on polyolefin oxidation in air.

References

[1] Patterson, G. H., and Riga, A. T., "Factors Affecting Oxidation Properties in Differential Scanning Calorimetric Studies," Thermochimica Acta, Vol. 226, 1993, pp. 201-210.

[2] Stricklin, P. L., Patterson, G. H., and Riga, A. T.," The Development of a Standard Method for Determining Oxidation Induction Times of Hydrocarbon Liquids by Pressure Differential Scanning Calorimetry," Thermochimica Acta, Vol. 243, 1994, pp. 201-208.

[3] Riga, A. T., and Patterson, G. H., "Standard Test Method for Determining Oxidative Induction Time of Hydrocarbons by DSC and PDSC," presented at ASTM Symposium, Nov. 21-22, 1996, New Orleans, LA.

[4] Riga, A., Young, D., Mlachak, G., and Kovach, P., "Thermoanalytical Evaluation of Readily Available Reference Polymers," Journal of Thermal Analysis, Vol. 49, 1997, pp425-435.

Inna Dolgopolsky,[1] Yuri I. Gudimenko,[2] and Jacob I.Kleiman[3]

THERMOANALYTICAL APPROACH FOR DIFFERENTIATION BETWEEN SURFACE AND BULK OXIDATION FOR POLYETHYLENE FILMS

REFERENCE: Dolgopolsky, I., Gudimenko, Y. I., and Kleiman, J. I., "**Thermoanalytical Approach for Differentiation Between Surface and Bulk Oxidation for Polyethylene Films,**" Oxidative Behaviour of Materials by Thermal Analytical Techniques, ASTM STP 1326, A.T. Riga and G. H. Patterson, Eds., American Society for Testing and Materials, 1997.

ABSTRACT: An analysis was done of changes in the crystallinity, the oxidation induction time (OIT) and the degree of surface oxidation of polyethylene thin films before and after gas-phase oxidative surface modification. Various surface treatments were performed which allowed to differentiate between oxidation of the bulk and the surface. It was shown that for a full characterization of oxidation processes in solid films all three parameters (degree of crystallinity, surface energy and OIT) needed to be taken into account. The correlations between the degree of crystallinity and OIT can be used as an indicator of the extent of bulk modification. The correlations between the surface energy and the OIT can be used as an indicator of surface oxidation.

KEYWORDS: polyethylene, gas-phase modification, surface energy, oxidation induction time, thermal analysis

[1] Manager, Thermal Analysis Division, Integrity Testing Laboratory Inc., 4925 Dufferin Street, North York, Ontario, M3H 5T6, Canada; Presently Senior Chemist, Woodbridge Foam Corp., 8214 Kipling Ave. North, Woodbridge, Ontario, Canada; tel: (905)-851-3914; fax: (905) 851-7294

[2] Senior Research Scientist, Integrity Testing Laboratory Inc., 4925 Dufferin Street, Downsview. Also University of Toronto Institute for Aerospace Studies, 4925 Dufferin Street, Downsview Ontario, M3H 5T6

[3] President, Integrity Testing Laboratory Inc., 4925 Dufferin Street, North York, Ontario, M3H 5T6, Canada; tel: (416) 667-7742; fax: (416) 667-7799; e-mail: jkleiman@utias.utoronto.ca; also Adjunct Professor, University of Toronto Institute for Aerospace Studies, 4925 Dufferin Street, Downsview Ontario, M3H 5T6 . To whom correspondence should be addressed.

The oxidative behaviour of polyolefins in general, and polyethylene (PE) in particular, is of great importance in many fields of polymer chemistry and different industrial processes. Despite the fact that polyolefins are chemically nonreactive, PE oxidizes and degrades under the influence of such factors as UV radiation, temperature, and the oxygen content of the air. Oxidative surface pretreatment of polyolefins is normally necessary in order to obtain satisfactory properties for adhesive bonding. At the same time polyolefins have low surface free energy and this limits their practical use. Among the numerous methods that can be used to improve surface properties of PE, the modification by corona discharge, photooxidation, ozonation, plasma, and combined UV/air or UV/ozone are of great importance [1,2].

A common peculiarity in the gas-phase surface oxidation processes is the generation and use of such reactive species as ozone, excited oxygen, singlet oxygen, thermal atomic oxygen, oxygen ions, sometimes in combination with UV and electron excitation of polymer molecules or additives in a polymer matrix. Numerous applications of polymers in space environment had triggered extensive investigations of polymer materials' behaviour, polyethylene among them, under hyperthermal atomic oxygen with kinetic energy of about 5 eV [3].

The main features of oxidation mechanisms and kinetics of polyolefins can be described by the theory of radical chain reactions with degenerated branching [4]. According to this theory, the oxidation depth of polymer films will depend on the structure of the polymer, the reactive species, and a few other conditions (to be discussed below). It is a well established opinion that corona discharge and plasma treatments oxidize only a fairly thin surface region (approximately 10 nm) [5], while the UV/air, ozone, or UV/ozone treatments affect a few micrometers of the surface [2]. The hyperthermal atomic oxygen initiates the erosion and changes of the upper layers of the polymer surface. The surface and bulk modification of polymers by gas phase oxidation treatment, and its control are still poorly understood, mainly due to the lack of basic and fundamental information. It is hoped that the results presented in this work will contribute toward such understanding.

EXPERIMENTAL METHOD

Test Procedure

The 30 μm thick polyethylene film, composed of a blend of low density and linear low density polyethylene (LDPE/LLDPE: 75/25) was supplied by Uniplast Industries Inc. (Orillia, Ontario). The studied PE exhibited two melting peaks at 111 °C and 120° C (Fig. 1).

Ozone treatment of PE was carried out at room temperature in a purged metal cell containing a quartz window. The ozone generated in dry air by a Simpson Industries Ltd.

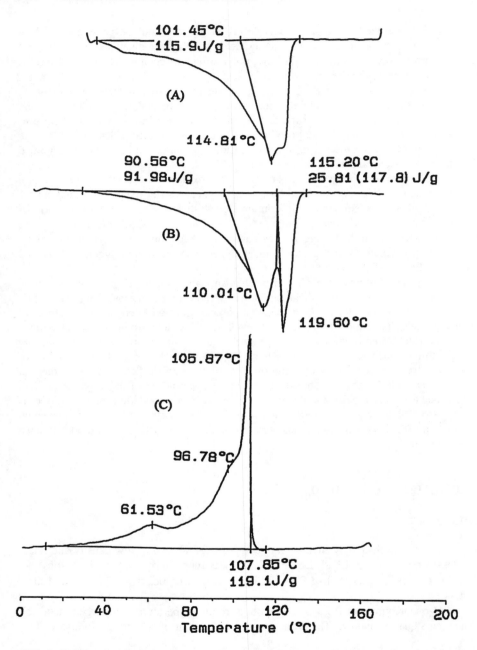

FIG. 1--DSC curve for pristine PE: (A) first heating, (B) second heating, (C) cooling.

generator. The gas flow was maintained at 0.42 L/min, and the ozone content was about 0.25 mol%. The same cell was used to expose the samples to a combination of ozone and ultraviolet radiation in a commercially available UV/Ozone Cleaning System (UVOCS) that generates UV emission in the 185 and 254 nm range (predominantly 254 nm). The ultraviolet intensity after the quartz window, which depends on the distance from the UV source, was 2.1 mW/cm² at a distance of 25 mm from UV bulbs. The photo-modification experiments in the air atmosphere were conducted in the UVOCS system, with UV intensity of 2.8 mW/cm² with the quartz glass removed.

All corona discharge experiments were carried out in ambient air, using a commercial system (Sherman Solid State Treater, model GX10). The system used an adjustable power output (0-1 kW), equipped with a high-frequency generator (5-8 kHz), high-power transformer, and a bench top Electrode Box Unit that consisted of two 1 cm by 30 cm discharge electrodes. The corona discharge takes place between the electrodes and the surface of the film. The travel velocity of the Electrode Box could be adjusted continuously. The distance between the polymer surface and the electrode was fixed at 2 mm. The relative humidity was 55-65%. The energy output was set at $E_u = 40$ mJ/mm², per unit of the substrate surface area. It was determined from a simple relationship

$$E_u = P/LV,$$

where,

P - power output (200 Watts),
L - the length of the treatment electrode (300 mm),
V - electrode box velocity (16.7 mm/s).

The atomic oxygen exposure tests were conducted at the University of Toronto Institute for Aerospace Studies Atomic Oxygen (AO) Beam Facility [6]. Each sample was placed in a holder that oriented the sample at 45° to the trajectory of the AO beam. The samples were exposed to hyperthermal atomic oxygen (HAO) with energy 3.0 eV at average flux of 1×10^{16} atoms/cm²·s for a period of 8 hours and were held at a constant temperature for the duration of the test. The areas of the samples under the holder, i.e. masked from the HAO, were considered as exposed to the thermal component of the atomic oxygen (TAO) with kinetic energy of approximate 0.04 eV.

Advancing contact angles between the polymer surface and droplets of specially selected liquids (deionized water and diiodmethylene (DIM)) were measured using 2-3 μl droplets. The surface-free energy (γ_s) and its dispersion (γ_s^d), and polar (γ_s^p) components were calculated for all samples from contact angles by the Harmonic-mean method [7].

Melting and crystallization behaviour of PE films were studied by Differential Scanning Calorimetry (DSC) using a DSC-10 Cell (TA Instruments). A controlled rate of cooling was provided by the Liquid Nitrogen Cooling Accessory (LNCA). Samples were tested in inert atmosphere (purge of nitrogen 50 ml/min). The samples were heated at 10°C/min up to 200 °C, held isothermally for 5 min, cooled down to 0 °C at 10 °C/min, held isothermally for 5 min and then heated at 10 °C/min up to 200 °C. The degree of crystallinity was estimated by using the enthalpy of melting for the second melting. The enthalpy of melting of a 100% crystalline polyethylene was assumed to be 293 J/g [8].

The Differential Scanning Calorimeter (DSC-10) with the Thermal Analyser

(Thermal Analyst 2000) were employed for oxidative induction time (OIT) determination, using aluminium sample holders. The temperature calibration of the instrument was done using indium in accordance with ASTM method E967-92. The indium sample was heated from 145 °C to 160 °C at 1 °C/min in nitrogen at 50 ml/min. The flow rates of gases were monitored using a "Matheson" flow meter SCCM AIR 600 (E100). The onset of indium melting and the maximum peak of indium melting were determined according to ASTM method E967. For calibration purposes, the temperature of indium melting was calculated at the extrapolated onset temperature. The resultant temperature (157.1 °C) was used as the calibration temperature. The reference temperature for indium melting is 156.6 °C.

The non-isothermal mode was used for determination of the ceiling temperature of PE oxidation. A PE film sample was heated at10 °C/min in an atmosphere of dry air (purge 50 ml/min). The onset of baseline increase (exothermically) was considered as the onset temperature of oxidation (Fig.2b). Two specimens were tested for each procedure. The specimens' weight was 5.65 $\pm$ 0.03 mg. The determination of OIT was done in the isothermal mode at a temperature lower then the onset of decomposition. The OIT scans were conducted in the following sequence: the specimen was placed in the holder and clamped, purged with nitrogen for 5 min prior to testing, heated in nitrogen at 10 °C/min to 189 °C then equilibrated and held isothermally at this temperature. Once the temperature was reached equilibrium at 189 °C, the nitrogen flow was changed to oxygen. This introduction of oxygen was considered as the beginning of the experiment ($\tau=0$). The isothermal operation was continued until the oxidative exotherm was observed and the procedure was stopped 2 min after the appearance of the oxidative exotherm.

RESULTS AND DISCUSSION

The ability of polymeric materials to preserve their structural, physical and other performance characteristics depends primarily on whether the oxidative modification processes are localized at the surface or develop uniformly in the bulk. The degree of oxidative modification development is affected strongly by reacting species' (molecules and radicals) transport from and to the surface. Migration of active species in solid polymers plays an important role in the degradation processes of macromolecules by transfering active species from one part of the polymeric material to another, thus causing molecular degradation in the bulk of the material.

The final goal of a surface modification process, which is a combination of chemical and physical changes occurring at the surface during the modification process, is to impart new, useful properties to the material. Complicated chemical processes arise during modification due to a variety of active oxygen species. These species, such as atoms, ions, and unstable molecules are responsible for generation of competing reactions of macromolecular oxidation, degradation and cross-linking. Despite the multiplicity of modification technologies and the use of different active species, the radical and the chain-radical mechanisms are the most important. These mechanisms play a leading role in photooxidative, corona discharge, ozone and atomic oxygen polymer surface modification processes. The oxidation of polyethylene can be assumed to proceed according to the general scheme involving initiation, propagation, chain branching, and termination [4].

Initiation:	$PH \rightarrow P\cdot + H\cdot$	(1)
Propagation:	$P\cdot + O_2 \rightarrow POO\cdot$	(2)
	$POO\cdot + PH \rightarrow POOH + P\cdot$	(3)
Chain Branching:	$POOH \rightarrow PO\cdot + \cdot OH$	(4)
	$PO\cdot + PH \rightarrow POH + P\cdot$	(5)
	$HO\cdot + PH \rightarrow H_2O + P\cdot$	(6)
Termination:		
	$2\ P\cdot$ ⎫	(7)
	$2\ POO\cdot$ ⎬ inactive products	(8)
	$P\cdot + POO\cdot$ ⎭	(9)

Scheme 1: Steps in an oxidation process of a polymer

The auto-oxidation process (in absence of hydroperoxides or free-radical initiators) generally involves an initiation step (1) in which a polymer macroradical or alkyl radical ($P\cdot$) is generated from the initial polymer by some means such as heat, light, etc. In this scheme PH denotes the polymer chain, with the H-being the most mobile hydrogen. The reaction of polymer (PH) with oxygen, to form radicals, can take place as well, namely, $PH + O_2 \rightarrow P\cdot + \cdot OOH$, where ($\cdot OOH$) - is a hydroperoxy radical.

Photooxidation

Photooxidation process in "pure" PE can be attributed to the presence of impurities, unsaturation, and to presence of oxidation products absorbing UV light [9]. It has been observed that photochemical oxidation of PE is localized within a few micrometers of the surface and decreases quickly with depth [10]. Changes in surface tension of PE films were used to monitor the degree of photooxidation of the surfaces in the present work. The photooxidation was associated with an increase in surface tension and with an increase in the polar component (γ_s^p) of the surface tension (Table 1) that is due to oxygen-containing functional groups. It should be noted that the dispersion component of surface tension (γ_s^d) practically didn't change. The PE is a semicrystalline polymer containing 40.6% of crystalline phase (Table 2). Oxidative processes affect predominantly the amorphous phase of the PE structure. The crystalline phase of PE is considered virtually unaffected, due to its relative impermeability to oxygen. The DSC analysis of UV/air exposed films (Table 2) had shown that the degree of PE crystallinity decreased, after 2 to 8 min exposure. The decrease in crystallinity can be a result of formation of oxidized groups in the bulk. After a 30 min exposure, the degree of crystallinity increased again. Such an increase can be explained by chain scission, which results in a higher mobility of shorter chains in the amorphous region that, in turn, gives rise to secondary or so-called oxidative crystallization [11]. For 30 μm thick PE films, UV/air irradiation, even for short periods, can lead to changes not just of surface properties but of bulk as well. Our experiments have demonstrated that these changes are strong enough to be detected by

Table 1--Summary data of polyethylene samples surface analyses

Sample	Contact angle, (degree)		Surface tension, (dyn/cm)		
	H₂O	DIM	γ_s^d	γ_s^p	γ_s
PE pristine	93	65	21.9	7.3	29.2
Photooxidation UV/air					
2 min	68	53	23.7	18.4	42.1
8 min	61	52	23.7	22.8	46.5
30 min	52	45	25.8	26.5	52.3
Ozonation					
15 min	100	52.5	34.0	1.5	35.5
30 min	97	56.6	28.5	3.7	32.2
60 min	91	54	27.6	6.2	33.8
Photoozonation UV/ozone					
2 min	62	46	26.1	21.0	47.1
8 min	60	39	28.9	21.0	49.9
Corona discharge treatment					
40 mJ/mm²	55	45	25.8	26.5	52.3
Atomic oxygen					
HAO	52	47	25.1	27.3	52.4
TAO	26	41	26.5	40.6	67.1

TABLE 2--Thermal Properties of Polyethylene Films Obtained from DSC analysis

Sample	T m$_1$, (°C)	ΔH$_1$, (J/g)	Tm$_2$, (°C)	ΔH$_2$, (J/g)	Degree of crystallinity (%)	OIT, (min)
PE pristine	114.7	115.6	110.9/120.0	120.3	40.64	13.9
Photooxidation in UV/air						
2 min	114.9	112.3	110.7/119.8	111.8	38.0	
8 min	116.0	87.9	110.8/119.9	92.2	31.3	10.5
30 min	116.0	124.8	113.0/120.1	123.9	41.9	
Ozonation						
15 min	115.5	109.3	110.9/120.4	107.9	36.7	22.4
30 min	114.5	111.8	111.0/120.3	110.1	37.2	
60 min	115.4	108.6	111.0/120.0	103.5	35.0	
Photoozonation UV/ozone						
2 min	115.6	120.8	112.1/120.7	115.3	38.9	
8 min	115.2	107.7	111.0/120.0	109.2	36.7	4.8
15 min	114.8	123.2	110.4/120.2	113.7	38.8	4.8
Corona discharge treatment						
40 mJ/mm^2	114.6	120.0	110.8/120.2	122.8	41.5	13.7
Atomic oxygen						
HAO	114.3	108.0	110.0/119.4	112.7	38.3	
TAO	114.2	114.3	109.8/119.3	115.9	39.4	

Tm$_1$ - melting temperature during first heating
Tm$_2$ - melting temperature during second heating
ΔH$_1$ - enthalpy of melting during first heating
ΔH$_2$ - enthalpy of melting during second heating

thermoanalytical methods. The extend of bulk modification seems to increase with increase in irradiation time.

The DSC analysis of pristine PE had shown that it is thermally stable in inert atmosphere for temperatures over 250 °C (Fig.2a), while an intensive oxidation starts at about 200°C in an oxygen environment (Fig.2b). The temperature of 189 °C was chosen for the OIT determination, since this temperature is close to the threshold of oxidation for our set-up, while still being below the onset temperature of decomposition for PE in air (T_{onset} = 240 °C) [10]. The OIT, generally, reflects the sensitivity of polymers to thermal oxidation. The latter is determined mainly by the chemical structure, concentration of components, and physical structural factors, such as chain conformation, crystallinity, supermolecular structure, geometry and degree of polymerization [12]. The increased thermal sensitivity of PE after an 8 min photooxidation process (OIT = 10.7 min), when compared with pristine PE (OIT = 14.8 min) (Table 2), can be attributed to the chemical changes in the polymer bulk. The photooxidation resulted in an increase in the concentration of hydroperoxide groups. A further process of hydroperoxide decomposition, for example, reaction 3, Scheme 1, accelerates the thermooxidation process of polymers [4], and thus, can decrease the OIT value. The oxidation isotherm for UV-treated PE (Fig. 3b) exhibited an anomalous exotherm effect after 3 minutes. This anomaly can be explained by the inhibiting effect of additives in PE films after photooxidation which extend the OIT time of PE during the thermal oxidation.

Ozonation and Photoozonation

Based on the experiments with ozonation of PE samples, the general trends in contact angles' behaviour and in the degree of crystallinity for ozonated and photoozonated samples were found to be similar to photooxidized PE samples. The increase of the polar component of surface tension in ozonation experiments with increased treatment time (Table 1) was accompanied by a decrease in the degree of crystallinity (Table 2) when compared to the pristine sample. It should be noted that under ozonation conditions used in this study, the occurred changes were found to be much lower when compared to photooxidation and photoozonation. As can be seen from Fig. 3c, the OIT parameter increased for samples exposed to ozone for 15 min. It should be noted that, generally, an increase in OIT was observed for all samples exposed to different times, up to 60 min. The observed effect can be explained by the specific mechanism of ozone reaction with double bonds of terminated groups in PE as well as by a reaction of additives, always present in PE films, with the ozone [13]. It is possible that more stable compounds result from these reactions.

The OIT time for PE after photoozonation (Fig. 3d) decreases as compared to OIT for pristine and photooxidized PE. This can be explained by the more severe conditions of photoozonation, in concert with the formation of singlet and atomic oxygen during ozone photolysis at 254 nm [14].

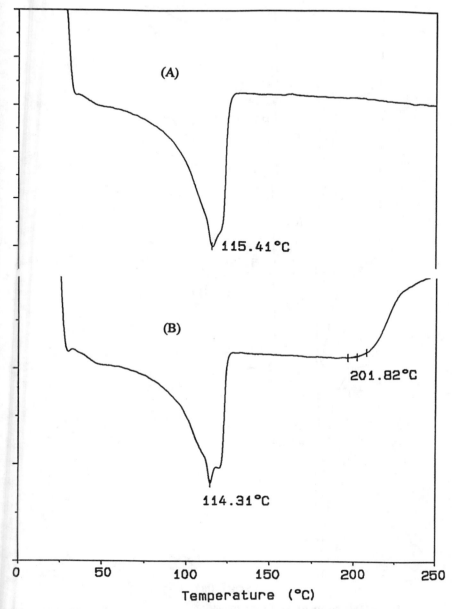

FIG. 2--DSC curves for pristine PE (heating) in atmosphere of nitrogen (A) and air (B)

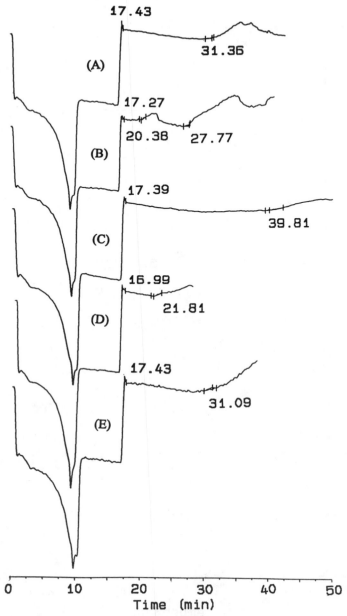

FIG. 3--DSC curve for OIT determination:
A - pristine PE, B- after UV/air exposure 8 min, C- after exposure 15 min,
D- after UV/ozone exposure 15min, and E - after corona discharge.

Corona Discharge and Atomic Oxygen Treatment

Based on XPS analysis, the depth of PE oxidation in the industrial corona discharge unit at $E_u = 1$ to $3\,mJ/mm^2$ was estimated at 5 to 7 nm [15]. In our experiments, the surface free energy of treated PE reached a value of $52.3\,mJ/m^2$ after corona treatment at $E_u = 40\,mJ/mm^2$. Despite a significant change in surface energy (Table 1), the degree of crystallinity, practically, didn't change (Table 2). As expected, the OIT, for corona discharge treated sample, was found to be the same as for the pristine sample (Table 2). This result indicated that functionalization of the surface took place, and that the bulk structure of PE remained unaffected under corona discharge, in a broad range of output energies. The functional groups, developed on the surface, consist, most probably, of relatively stable oxygen-containing functionalities [16] which are not involved in initiation of bulk oxidation. These results support the suggestion that the OIT parameter may be used for characterization of oxidative behaviour of polymer materials under different conditions of gas-phase surface modification.

The principal changes in a polymer exposed to atomic oxygen are in the mass loss [6, 17], the chemical composition of the surface, and in molecular weight of the surface layer [18]. It was assumed, traditionally, that the effects of atomic oxygen treatment are confined to a thin layer of the surface, and that the bulk properties of treated polymers remain unchanged [19]. For PE exposed to HAO under various experimental conditions, a decrease in crystallinity [20], and changes of surface morphology [21] were observed.

Preliminary analysis of PE exposed to a moderate fluence of atomic oxygen (2.9×10^{20} atoms/cm^2) indicated that the degree of crystallinity of the exposed PE decreased when compared to the control PE. This tendency is less pronounced for PE exposed to HAO (Table 2). Due to the geometry of the sample no reliable measurements of the OIT parameter were made. Work is now in progress on HAO exposed samples to reduce this data.

CONCLUSION

It was shown that the differences in the types of oxidative treatments can be held accountable for the different behaviour of the OIT parameter and the degree of crystallinity, as observed in the analysed samples. Corona discharge treatment, despite strongly affecting the surface properties of the PE samples, had very little effect on the OIT and the degree of crystallinity. These observations are well in accord with the accepted notion that corona discharge treatment is, indeed, a technique that affects mainly the surface. On the other hand, the photooxidation and the photoozonation treatments effected very strongly the OIT and the degree of crystallinity (both decreased) of the treated PE samples, which suggests a strong effect on the bulk of the material. In ozonation experiments with PE no clear trends were found. While the degree of crystallinity decreased with time, the OIT parameter increased, indicating that the interaction of ozone with polyethylene is different from other treatments.

More work is required to understand better the observed relationships and their dependence on the type of treatment. It would be very important, for instance, to study the influence of various antioxidants, the thickness of the treated films and the type of polyolefins on the observed relationships.

ACKNOWLEDGMENTS

The authors would like to acknowledge Dr. D. Morison for atomic oxygen experiments.

REFERENCES

[1]   Dasgupta, S., "Surface Modification of Polyolefins for Hydrophilicity and Bondability: Ozonization and Grafting Hydrophilic Monomers on Ozonized Polyolefins," Journal of Applied Polymer Science, Vol. 41, 1990, pp 233-248.

[2]   Strobel, M., et al., "A Comparison of Gas-Phase Methods of Modifying Polymer Surfaces,"Journal of Adhesion Science and Technology, Vol. 9, No. 3, 1995, pp 365-383.

[3]   Iskanderova, Z. A., et al., "Influence of Content and Structure of Hydrocarbon Polymers on Erosion by Atomic Oxygen," Journal of Spacecraft and Rockets, Vol. 32, No. 4, 1995, pp 878-884.

[4]   Emanuel, N. M. and Buchchenko, A. J., "Chemical Physics of Polymer Degradation and Stabilization, " VNU Science Press, 1987.

[5]   Hollander, A., Klemberc-Sapiena, J.E., and Wertheimer, M. R., "Induced Oxidation   of the Polymers Polyethylene and Polypropylene," Journal of Polymer Science: Part A: Polymer Chemistry , Vol. 33, 1995, pp 2013-2025.

[6]   Tennyson, R. C., "Atomic Oxygen Effects on Polymer Based Materials," Canadian Journal of Phys., Vol. 69, 1991, pp 1190-1208.

[7]   Wu, S., "Polymer Interface and Adhesion," Marcel Dekker, New York and Basel, 1982.

[8]   Wunderlich, B. "Macromolecular physics", Vol.3, Academic Press, New York, 1980.

[9]   Gugumus, F., "Photo-oxidation and Stabilisation of Polyethylene, "Mechanisms of Polymer Degradation and Stabilisation, G. Scott, Ed., Elsevier Applied Science, 1990, pp 169-210.

[10]   Qureshi, F.S., et al., "Weather-Induced Degradation of Liner Low-Density Polyethylene: Mechanical Properties," Polymer-Plastics Technology and Engineering, Vol. 28, No. 7&8, 1989, pp 649-662.

[11] Liu, M., Horrocks, A. R., and Hall, M. E., "Correlation of Physicochemical Changes in UV-exposed Low Density Polyethylene Films Containing Various UV Stabilizers,"Polymer Degradation and Stability, Vol. 49,1995, pp 151-161.

[12] Iring, M., and Tudos, F., "Thermal Oxidation of Polyethylene and Polypropylene: Effects of Chemical Structure and Reaction Conditions on the Oxidation Process," Progress in Polymer Science, Vol. 15, 1990, pp 217-262.

[13] Peeling, J., "Surface Ozonation and Photooxidation of Polyethylene Film," Journal of Polymer Science: Polymer Chemistry Edition, Vol. 21, 1983, 2047-2055.

[14] Rabek, J. F., et al., "Photoozonization of Polypropylene. Oxidative Reactions Caused by Ozone and Atomic Oxygen on Polymer Surfaces," Chemical Reactions on Polymers, J. L. Benham and J. F. Kinstle, Eds., American Chemical Society, Washington, DC, 1988, pp 188-355.

[15] Suntherland, I., Popat, R.P., and Brewis, D.M., "Corona Discharge Treatment of Polyolefins,"The Journal of Adhesion, Vol. 46, 1994, pp 79-83.

[16] Briggs, D., and Kendall, C. R., "Derivatization of Discharge-Treated LDPE: an extension of XPS Analysis and a Probe of Specific Interactions in Adhesion," International Journal of Adhesion and Adhesives, January 1982, pp 13-17.

[17] Golub, M.A., and Wydeven, T., "Reactions of Atomic Oxygen (O(^{3}P)) with Various Polymer Film," Polymer Degradation and Stability, Vol. 22, 1988, pp 325-338.

[18] Golub, M.A., and Wydeven, T., "CSCA Study of Kapton Exposed to Atomic Oxygen in Low Earth Orbit or Downstream from a Radio-Frequency Oxygen Plasma,"Polymer Communications, Vol. 29. 1988. pp 285-288.

[19] Golub, M. A., Lerner, N. R., and Wydeven, T., "Reaction of Atomic Oxygen [O(^{3}P)] with Polybutadienes and Related Polymers," American Chemical Society Symposium Series, Vol. 364, 1988, pp 342-355.

[20] Strganac T. W., et al., "Characterization of Polymer Films Retrieved from NASA's Long Duration Exposure Facility" Journal of Spacecraft and Rockets, Vol. 32, No. 3, May-June 1995, pp 502-506.

[21] Kleiman, J., et al., "Surface Treatment of Materials and Structure by Hyperthermal Atomic Oxygen," Proceedings of the 8th CASI Conference on Astronautics, November 8-10, pp 451-461, Ottawa, Canada, 1994.

Clarence J. Wolf,[1] Scott C. Hager,[1] and Nick P. Depke[1]

THERMO-OXIDATIVE DEGRADATION OF IRRADIATED ETHYLENE TETRAFLUOROETHYLENE*

REFERENCE: Wolf, C. J., Hager, S. C., and Depke, N. P., **"Thermo-Oxidative Degradation of Irradiated Ethylene Tetrafluoroethylene,"** Oxidative Behavior of Materials by Thermal Analytical Techniques, ASTM STP 1326, A. T. Riga and G. H. Patterson, Eds., American Society for Testing and Materials, 1997.

ABSTRACT: The thermal oxidative degradation of the semicrystalline thermoplastic polymer, ethylene tetrafluoroethylene (ETFE), used for electrical insulation was studied as a function of radiation (1.5 MeV electrons), temperature, and contact with a metal surface. The radiation, whose total dose varied from 0 to 48 MRads, produces two effects: (1) cross-linking which enhances the high-temperature mechanical properties of the polymer and (2) degradation which reduces its high-temperature lifetime. Modified wires were constructed by replacing the original silver-plated copper conductor with tin-plated copper to ascertain the effect of metal surface on the degradation process. The rates of degradation of insulation (no conductor), original wires, and modified wires were measured by weight loss and by thermogravimetric analysis (TGA). The primary degradation products were identified by evolved gas mass spectrometry. A simple model in which the rate of decomposition as a function of temperature, degree of conversion, and radiation dose was developed. The lifetime of the insulation in a thermal oxidative environment is a strong function of the composition of the electrical conductor it surrounds. Silver over copper was a much more effective catalyst in promoting the thermal oxidative degradation of the ETFE than tin over copper. The rate of degradation was a strong function of the total radiation dose regardless of conductor.

KEYWORDS: electrical insulation, ethylene tetrafluoro ethylene, thermal oxidative stability, aging, lifetime, radiation, catalyst

Modern high-performance aerospace systems have evolved into large platforms carrying highly sophisticated electronic equipment which must survive an extremely harsh environment. Many thousands of feet of electrical wiring are required to connect the various components of the system. Even relatively small fighter aircraft such as the Navy's F-18 Hornet requires more than 125 000 ft (38 100 m) of electrical wiring. The large amount of material coupled with the total energy requirements place a severe limit on the

[1]Materials Science and Engineering, Washington University, St. Louis, MO 63130.

*Supported by the Air Force Office of Scientific Research, Washington, DC.

overall properties of the wire: it must be low volume and lightweight, the insulation must exhibit excellent dielectric properties, and it must be thermally stable at high temperatures for long periods of time.

One material of particular interest for high-temperature insulation is ethylene tetrafluoroethylene (ETFE). ETFE is a semicrystalline thermoplastic resin that has excellent dielectric properties and melts in the range of 260 to 280°C. However, at temperatures approaching the melting point, mechanical properties, such as tensile strength, decrease an order of magnitude. To enhance the mechanical properties of ETFE at temperatures near 200°C, the resin is cross-linked. Cross-linking is usually accomplished by irradiating the ETFE in the presence of a reactive cross-linking agent. Although the radiation enhances the mechanical properties it has several undesirable effects: (1) it degrades the polymer and (2) it creates active species, such as peroxides or hydroperoxides or both, which decompose at elevated temperatures and severely limit the lifetime of the insulation.

We have investigated the thermal degradation of irradiated ETFE both in the presence and absence of the electrical conductor used in the manufacturer of electrical wiring by varying both the number of stands in the conductor and the surface plating. The effect of radiation dose, varying from 0 to 48 MRads, on the degradation rates was also determined.

EXPERIMENTAL PROCEDURE

The terminology used throughout this report is as follows: insulation refers to the polymer alone, conductor refers to the current carrying metal (either silver-plated or tin-plated copper), and wire refers to the actual insulated conductor. For those studies involving the insulation alone, it was carefully separated from the conductor with a precision wire stripping tool and used directly.

Materials

The ETFE was commercial grade polymer containing a cross-linking agent (triallyl isocyanurate). The wire, single-walled 22 gauge, was prepared by extrusion and irradiated with 1.5 MeV elections to the specified dose. The nominal conductor was 19-strand silver-plated copper. The diameter of each strand is approximately 0.15 mm and is plated with a layer of either silver or tin 5×10^{-4} mm thick. We constructed a "modified" wire by replacing the initial 19-strand conductor with few strands of either the silver/copper conductor or a conductor of identical size with tin plating rather than silver. The fraction of insulation on the wires was measured for all the irradiated samples and used to correct raw weight loss data to weight loss of insulation. Thus, in all cases, weight loss data refers to weight loss of insulation. The conductor itself exhibited a slight change in weight during isothermal aging at 300°C: after 30 min at 300°C the weight decreased approximately 0.04% followed by a slow rise which reached 100.04% of the original weight after 8 h.

The modified wires used in this study are summarized in Table 1. The geometrical contact area, given in terms of square centimeters per length, between the conductor and the insulation are given in Table 1.

Thermo-oxidative Aging

Selected samples, approximately 30 cm long, were aged for a specified time at temperatures ranging from 230 to 300°C in a Precision Scientific Mechanical Convection Oven 625. All samples were run in triplicate. The weight of the individual samples was measured at selected times and a weight loss versus time curve was established for all wires and insulation.

TABLE 1--<u>Modified wire systems studied.</u>

Plating	Number of Strands	Contact Area $\times 10^{-2}$ cm^2/L	Radiation Dose MRads			
Ag	0	0	0	9	29	48
Ag	5	8.3	0	9	29	48
Ag	9	11.6	0	9	29	48
Ag	13	16.7	0	9	29	48
Ag	19	21.6	0	9	29	48
Sn	0	0	0	9	29	48
Sn	5	8.3	0	9	29	48
Sn	9	11.6	0	9	29	48
Sn	13	16.7	0	9	29	48

Thermogravimetric Analysis

A TA-2910 TGA equipped with automatic sample inlet was used for all TGA analysis. The degradations were conducted in air at a flow rate of 60 cm^3/min. Both isothermal and the thermal jump method were used to evaluate degradation rates. In the latter method, the temperature is "jumped" from T_1 to T_2; thus rates can be measured at two temperatures at the same degree of conversion to determine kinetic parameters directly [1].

RESULTS AND DISCUSSION

The weight loss curves from a series of insulations and wires (Ag/Cu conductor) irradiated to total doses varying from 0 to 48 MRads as a function of time at 240°C are shown in Fig. 1. The open data points refer to wire while the solid points correspond to the insulation. For each sample, weight loss is approximately linear with time (that is, constant rate of degradation). Two important observations are to be noted: 1) the weight loss increases rapidly with total radiation and 2) the rate of degradation of wire is significantly greater than that of the insulation. The primary volatile degradation products are HF and low molecular weight hydrofluorocarbons of the type [CH$_2$CH$_2$CF$_2$CF$_2$]$_n$ [2]

Typical rate of degradation curves (TGA derivative) at 300°C from the insulation irradiated at 9, 19, 29, and 39 MRads are shown in Fig. 2. The rates vary a factor of 2 from 0.008 to 0.015 %/min for the insulation irradiated at 9 and 39 MRads, respectively. A comparison of isothermal rates and rates obtained by the "jump-method" are shown in Fig. 3. Isothermal runs at 310 and 300°C are shown together with the jump run in which the temperature was cycled between 310 and 300°C. The temperature was decreased from 310 to 300° at weight losses corresponding to 3, 5, and 7% and the temperature increased from 300 to 310°C at weight losses corresponding to 4, 6, and 8%. Although kinetic parameters can be determined at all of these weight loses, the values at 3 and 4% are discarded because weight loss of residual low molecular weight material cannot be distinguished from that as a result of thermal degradation.

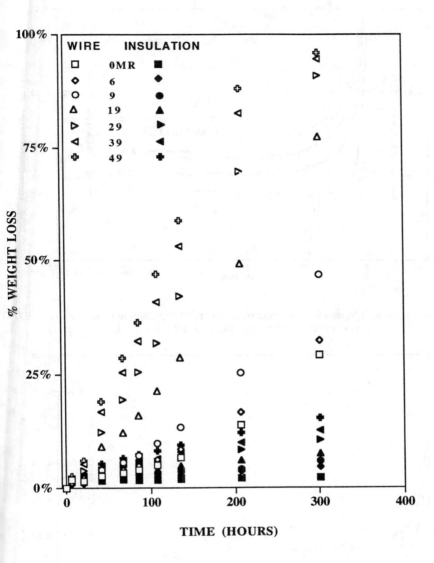

FIG. 1 – Weight loss as a function of time for the isothermal oxidative degradation of irradiated insulation and wires at 240°C.

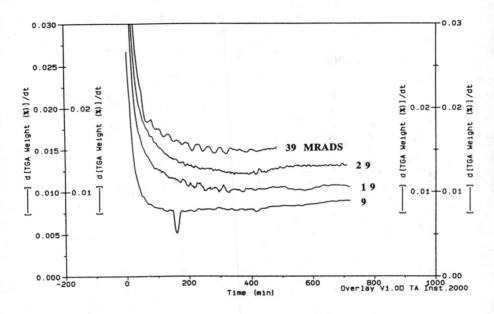

FIG. 2 – Rate of weight loss for the isothermal oxidative degradation of insulation at 300°C as a function of time for different radiation levels.

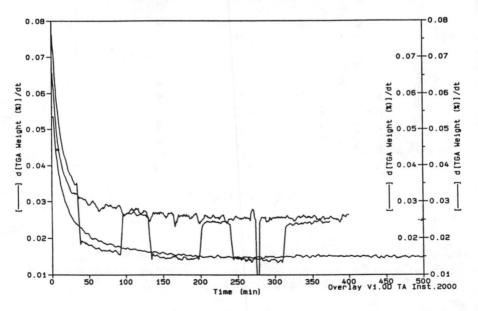

FIG. 3 – Comparison of the rate of degradation of insulation by isothermal analysis and the thermal jump method (at 300 and 310°C).

The degradation rates as a function of radiation dose for 5% weight loss from the insulation and wire at 300 and 310°C are shown in Figs. 4a and 4b, respectively. The rate increases linearly with dose. The rate of change, that is, the slope of the degradation rate versus radiation dose curve for both insulation and wire at 300 and 310° C are summarized in Table 2. Temperature coefficients, determined from a classical Arrhenius plot of the logarithm of the rate of weight loss as a function of reciprocal temperature, are also listed in Table 2. The lower temperature coefficient observed for the wire is consistent with the observations noted in Fig. 1, that is, higher rate of degradation in wire compared to insulation.

The effect of tin-plated copper compared to silver-plated copper is illustrated in Fig. 5. The rate of degradation at 300°C in modified wire systems containing 0 to 19 strands of the plated conductor in an unirradiated construction at 5% total weight loss is shown in Fig. 5. The rate of weight loss increases linearly in the Ag/Cu system, at a rate of approximately 1×10^{-5} wt%/min per cm^2 of contact. The rate of degradation from the tin-plated copper system is approximately 0.2 wt%/min but is independent of the contact area between the plated conductor and the insulation. For a 19-strand system, this corresponds to a four-fold increased degradation rate in the silver-based compared to the tin-based construction.

If we assume that the degradation process can be modeled as a series of independent events, the rate of degradation R is given by

$$R = R(T) \times R(\alpha) \times R(D) \qquad (1)$$

where the individual rates as a result of to temperature, degree of conversion, and radiation dose are given by $R(T)$, $R(\alpha)$, and $R(RD)$, respectively. The individual terms are:

$$R(T) = R_o(T)\exp(-E/RT) \qquad (2)$$

$$R(\alpha) = R_o^{'}(\alpha)[1-\alpha]^n \qquad (3)$$

$$R(D) = R_o^{''}(D)^m \qquad (4)$$

where R_o, $R_o^{'}$, and R'' are the appropriate proportionality factors relating temperature, degree of conversion, and radiation dose to rate, respectively.

The constants for these three equations are evaluated for both the insulation alone and from the wire. The dependence of the rate on the appropriate variable is determined from the slope of a log-log plot. (The temperature terms in Eq 2 are determined from a classical Arrhenius plot.) Equations 3 and 4 can be rewritten as

$$\log[R(\alpha)] = \log R_o^{'} + n\log(1-\alpha) \qquad (5)$$

$$\log[R(D)] = \log R_o^{''} + m\log(D) \qquad (6)$$

The rate of degradation (at 300°C) as a function of degree of conversion for the radiation doses from 0 to 48 MRads plotted in logarithmic form (Eq 5) are summarized in Figs. 6a and 6b for the insulation and wire (containing 19 strands of Ag/Cu conductor), respectively. The average value for the slopes for doses from 6 to 48 MRads in the 5 to 8% decomposition range are 1.38 and 1.88 for the insulation and wire, respectively. The logarithm rate of degradation (at 300°C) as a function of the logarithm of the radiation dose for samples degraded to 5, 6, 7, and 8% are shown in Figs 7a and 7b for the insulation and wire, respectively. For the insulation, the rate can be conveniently divided into two regions

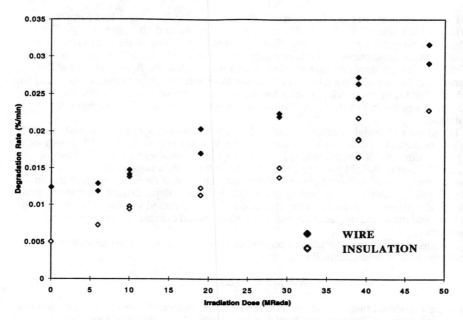

FIG. 4a – Rate of degradation at 5% weight loss for insulation and wire as function of radiation dose at 300°C.

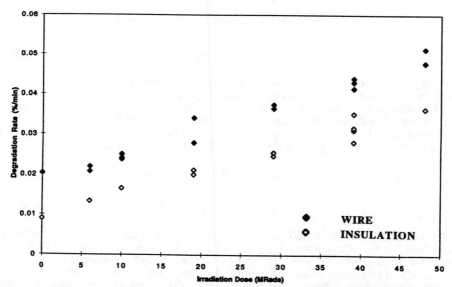

FIG. 4b – Rate of degradation at 5% weight loss for insulation and wire as function of radiation dose at 310°C.

TABLE 2--<u>Change in rate of degradation as a function of radiation dose in insulation and wire (Ag/Cu conductor).</u>

Materials	Temperature °C	Rate of Change x10⁻⁴ wt%/MRad	Temperature Coef. kJ/mol
Insulation	300	3.5	
Insulation	310	5.7	135
Wire	300	4.1	
Wire	310	6.4	124

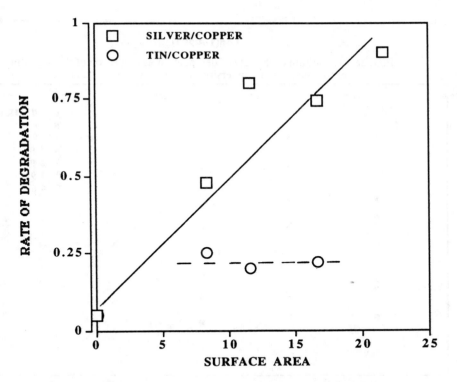

FIG. 5 – Rate of degradation of modified wires at 300°C as a function of the contact area between the conductor and the polymer.

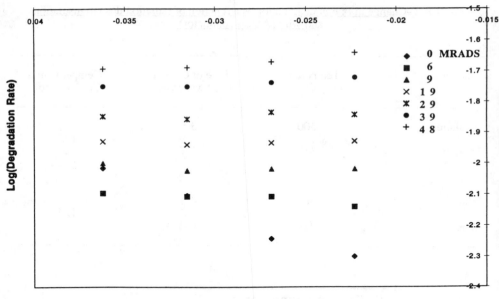

FIG. 6a – Logarithm of the rate of degradation as a function of the logarithm of (one-degree of conversion) for the irradiated insulation.

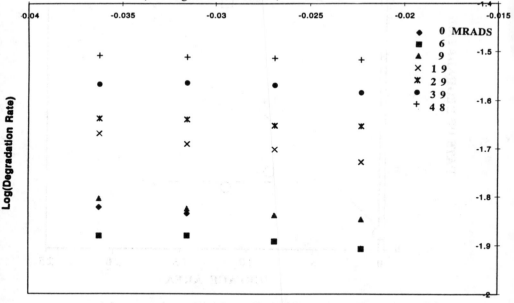

FIG. 6b – Logarithm of the rate of degradation as a function of the logarithm of (one-degree of conversion) for the irradiated wire.

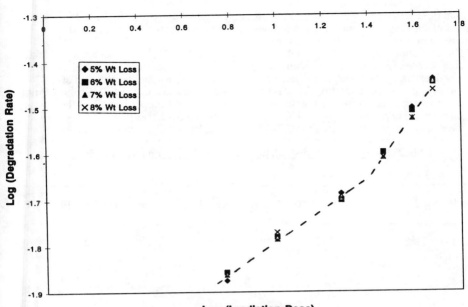

FIG. 7a – Logarithm of the rate of degradation of the insulation as a function of the logarithm of the dose at 5, 6, 7, and 8% conversion.

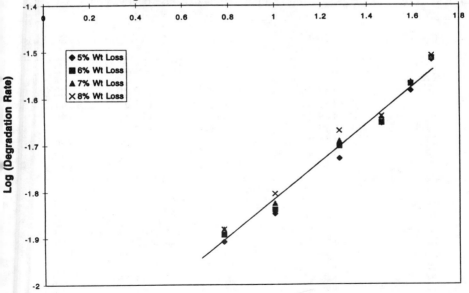

FIG. 7b – Logarithm of the rate of degradation of the wire as a function of the logarithm of the dose at 5, 6, 7, and 8% conversion.

for total dose less than 29 MRads and the other for doses greater than 29 MRads. The slope is essentially constant for degraded wire and equal to 0.40. The slopes in these regions (m in Eq 6) are briefly summarized in Table 3.

The activation energies determined from Arrhenius plots of the rate data are summarizes in Table 4. Averaged values and standard deviations for each set of values are also listed. These values are again averaged to yield an activation energy for the degradation of the insulation and wire, E(insulation) = 141.2 ± 2.6 kJ/mol and E(wire) = 150.8 ± 4.0 kJ/mol

Other workers have reported that metals such as silver and copper can affect the degradation of polymers. Hong and Wang [3] studied the oxidative degradation of a brominated epoxy resin with FTIR and TGA. They added small amounts of CuO and

TABLE 3--<u>Summary of degradation rate dependence on radiation dose.</u>

Sample	Total Dose Received	m (slope from Fig. 7)
Insulation	< 29 MRads	0.36
Insulation	>29	0.84
Wire	0-48	0.40

TABLE 4--<u>Activation energies (kJ/mol) for the thermal oxidative degradation of X-ETFE.</u>

Radiation Dose MRads	5% Wt Loss	6% Wt Loss	7% Wt Loss	8% Wt Loss
Wire				
0	137.24	117.04	137.89	122.63
6	151.88	147.23	141.27	147.18
10	149.58	148.05	147.10	140.85
19	140.64	137.00	143.50	137.10
29	141.25	142.60	137.07	136.16
39	138.34	132.85	136.39	139.05
48	136.02	137.92	138.93	139.32
Ave	143 ± 6.4	140.9 ± 6.0	140.7 ± 4.1	140.0 ± 3.9
Insulation				
0	161.16	152.36	151.74	129.45
6	169.62	160.19	156.37	153.51
10	151.92	151.92	152.38	146.96
19	154.66	149.00	155.71	152.77
29	155.01	150.78	158.03	151.91
39	140.45	146.67	144.16	143.78
48	128.96	142.91	154.23	146.94
Ave	150.1 ± 13.9	150.3 ± 5.9	153.5 ± 4.9	149.3 ± 3.1

Cu_2O and noted that the copper oxides catalyzed the degradation process. They reported that the activation energy decreases when either of the oxides is added, however Cu_2O had a more pronounced effect than does CuO. Ulrich [4] et al. used TGA to investigate the thermal degradation of five polymer systems containing silver flakes. They noted that the temperature at which thermal degradation begins is 20 to 70°C lower when the material contains silver than without it.

CONCLUSION

The rate of degradation of cross-linked ETFE (XL-ETFE) was measured as a function of temperature, degree of conversion, radiation dose used to cross-link the polymer, and the effect of the conductor on the degradation rate. The standard silver-plated copper conductor accelerated the rate of thermal oxidative degradation compared to either the base polymer or to a tin-plated copper conductor. A simple model based on independent reactions was evaluated and numerical parameters determined. The rate of degradation of the insulation alone [R(insulation)] and insulation on the wire [R(wire)] are given below.

$$R(insulation) = R_o \exp\left(\frac{-16983}{T}\right) \times (1-\alpha)^{1.38} \times (Dose < 29MR)^{0.36}$$

$$= R_o \exp\left(\frac{-16983}{T}\right) \times (1-\alpha)^{1.38} \times (Dose > 29MR)^{0.84}$$

$$R(wire) = R_o \exp\left(\frac{-18138}{T}\right) \times (1-\alpha)^{1.88} \times (Dose)^{0.40}$$

where T is the temperature (K), α is the degree of conversion, and dose is in MRads.

REFERENCES

[1] Flynn, J. H., "Thermogravimetric Analysis and Differential Thermal Analysis" in Aspects of Degradation and Stabilization of Polymers, H. H. G. Jellinek, Ed., Elsevier, 1978, Chapt. 12, pp. 573-603.
[2] Morelli, J. J., Fry, C. G., Grayson, M. A., Lind, A. C., and Wolf, C. J., "The Thermal Oxidative Degradation of an Ethylene-Tetrafluoroethylene-Copolymer-Based Electrical Wire Insulation," Journal of Applied Polymer Science, Vol. 43, 1991, pp. 601-611.
[3] S. A. G. Hong and T. C. Wang, "Effect of Copper Oxides on the thermal Oxidative Degradation of Epoxy Resin," Journal of Applied Polymer Science, Vol. 52, 1994, pp. 1339-1352.
[4] M. Ulrich, L. Sarrof and Loucheux, "Thermal Analysis of Silver Coating Using for Tantalum Copicutors," Journal of Applied Polymer Science, Vol. 52, 1994, pp 387-399.

Mark F. Fleszar[1]

THERMAL OXIDATIVE STABILITY OF FIBER-REINFORCED EPOXY COMPOSITE MATERIALS

REFERENCE: Fleszar, M. F., "**Thermal Oxidative Stability of Fiber Reinforced Epoxy Composite Materials**", Oxidative Behavior of Materials by Thermalanalytical Techniques, ASTM STP 1326, A. T. Riga and G. H. Patterson, Eds., American Society for Testing and Materials, 1997.

ABSTRACT: The thermal oxidative stability of Fiberite 7714A (glass/epoxy), 976 (carbon/epoxy) and 977-2 (carbon/epoxy) were studied at temperatures up to and above their respective glass transition temperatures. Composite samples were subjected to both long and short durations of heating. The first set of samples were heated for 4 to 6 hrs from 100 to 350°C and the second set of samples were cycled for 30 min intervals from 200 to 400°C.

The samples were tested in a Perkin-Elmer TGA7 Thermogravimetric Analyzer and the weight loss recorded. The weight loss data was compared on the basis of temperature and time at temperature. The results showed good thermal stability for each epoxy material through its glass transition temperature with a weight loss of 1.5% or less. The results were comparable for both long and short-term temperature exposure.

KEYWORDS: epoxy, composite, thermal analysis, thermal stability, glass fibers, carbon fibers

[1]Chemist, Benet Labs, Watervliet Arsenal, Watervliet, N.Y. 12189

INTRODUCTION

Projectile launchers afford a unique application for composite materials. The introduction of organic composite materials can be used to increase dimensional stability and stiffness. Polymer blends can increase impact strength and thermal stability. Unlike other high temperature applications, cannon can see both sustained and short-term heating at elevated temperatures. They are frequently fired for relatively short periods of time, going through repeated cyclic heating. Given the performance requirements for present and future projectile launchers, a true service temperature for composite materials for each application should be established. Manufacturer specifications generally recommend a maximum service temperature for an extended period of time, ignoring any effect experienced by short-term cyclic heating. Depending on the composite system, cannon requirement for a minimum service temperature of 300°C may exceed the recommended service temperature.

Development of composite structures for use in projectile launchers requires knowledge of thermal degradation of composite materials after continual thermal cycling. Materials thought to be adequate, may or may not be satisfactory for the thermal application of projectile launchers. Once the thermal stability of a material is established, the effect of degradation on the mechanical properties must be established to determine a true service temperature.

EXPERIMENTAL

Test specimens were prepared from three Fiberite resin systems [7714A (glass/epoxy), 976 (carbon/epoxy) and 977-2 (carbon/epoxy)] as specified in the manufacture's product guide [1]. The specimens were sectioned and analyzed by a Perkin-Elmer Differential Scanning Calorimeter, Model DSC7, to determine their glass transition temperature (Tg). Samples of 10 to 15 mg were sealed in standard aluminum pans and heated in a nitrogen atmosphere at a scanning rate of 20°C/min. The Tg was determined by extending the pre- and post-transition baseline and taking the midpoint of the transition region [2].

The thermal stability of each composite material was evaluated in a Perkin-Elmer Thermogravimetric Analyzer, Model TGA7, using a 15 to 20 mg sample heated in an oxygen atmosphere. Two methods of isothermal analysis were performed to simulate different firing scenarios [2,3]. First, a set of samples were heated at a rate of 20°C/min to a specific testing temperature, then held at constant temperature, in the case of 7714A for 4 hrs at 100, 150, 200, 250 and 275°C, while 976 and 977-2 were heated for 6 hrs at 200, 250, 300, 325 and 350°C. Next, sample sets were heated at a rate of 20°C/min, then held at constant temperature at 200, 225, 250 and 300° C for 30 min, cooled at a rate of 20°C/min to 25°C and the procedure was repeated for a total of twelve cycles. In each case, the weight loss was measured for each cycle and temperature.

RESULTS and DISCUSSION

All three Fiberite prepreg materials analyzed were epoxy resins, reinforced with either glass or carbon fibers [1] (Table 1). Fiberite 7714A, a brominated epoxy resin with glass fibers, is a flame resistant polymer with a recommended service temperature of 71°C. The samples had an average

Table 1: Material properties

RESIN	PREPREG	RECOMMENDED SERV TEMP °C	Tg °C	AVE RESIN CONTENT %
7714A	EPOXY/ GLASS	71	120	27.4
976	EPOXY/ CARBON	177	212	37.5
977-2	EPOXY/ CARBON	174	174	35.8

resin content of 27.4% and a glass transition temperature (Tg) of 120°C. Fiberite 976 is an epoxy prepreg reinforced with carbon fibers; its recommended service temperature is 177°C. TGA analysis revealed an average resin content of 37.5% and a Tg of 212°C. Fiberite 977-2 is a toughened epoxy/carbon prepreg with a recommended service temperature of 149°C. The prepreg had an average resin content of 35.8% and a Tg of 174°C.

Fiberite 7714A, when heated in an oxygen environment for 4 hrs at constant temperatures up to 150°C showed good thermal stability with a weight loss of 0.35% (Table 2 and Fig. 1). The observed weight loss is most likely the result of a loss in absorbed water. When the temperature was increased from 150 to 200°C the weight loss increased to 1.08%, which indicates an onset of thermal decomposition of the sample. When the temperature was increased beyond 200°C, the decomposition of just the resin, not the glass fibers, increased dramatically, indicating severe material decomposition.

As seen in Fig. 2 and Table 2, when Fiberite 976 was heated in oxygen at a constant temperature of 150°C for 6 hrs, the weight loss was approximately the same as that observed in Fiberite 7714A, which had been heated for 4 hrs. The weight loss of Fiberite 977-2 was 0.71%, twice that

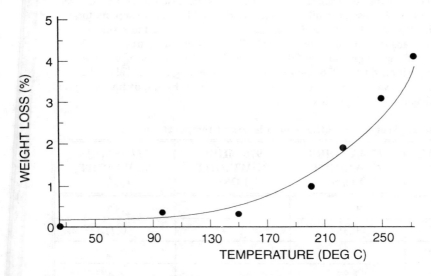

Figure 1. Plot of weight loss for Fiberite 7714A after 4 hrs versus temperature °C

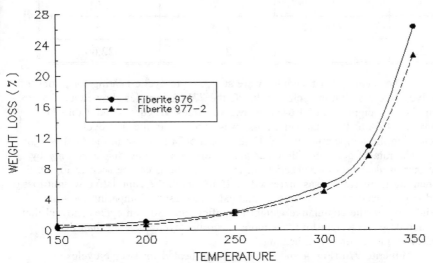

Figure 2. Plot of weight loss for Fiberite 976 and 977-2 after 6 hrs versus temperature°C

observed in 976, but when the temperature was increased to 200°C, its weight loss remained constant, while 976 increased to 1.11%. This supports the premiss that the observed weight losses up to 150°C were the result of moisture absorption and in the case of 977-2 was probably true up to 200°C. Increasing the temperature to 200°C again showed the onset of thermal decomposition for the 976 resin with 977-2 remaining stable. When the temperature was increased above 200°C, the rate of decomposition increased significantly for both resins.

Table 2:   Weight loss after 4 or 6 hours at temperature

TEMP °C	7714A 4HRS % WEIGHT LOSS	976 6HRS % WEIGHT LOSS	977-2 6HRS % WEIGHT LOSS
100	0.35	NA	NA
150	0.35	0.34	0.71
200	1.08	1.11	0.72
225	1.87	NA	NA
250	3.29	2.36	2.16
275	4.28	NA	NA
300	NA	5.76	5.01
325	NA	10.85	9.61
350	NA	26.34	22.60

The second set of samples were subjected to cyclic heating for a total of twelve cycles at 30 min/cycle. At 200°C the 7714A samples showed a cumulative weight loss of 1.64% for twelve cycles (Figure 3 and Table 3) however, 46% of the total weight loss was observed in the first cycle. Increasing the temperature to 225°C showed a 60% increase in the total weight loss. The rate of decomposition was almost the same, being displaced only by the increased weight loss observed in the first cycle as can be seen in Figure 3. When the temperature was increased to 250°C, both the cumulative weight loss and rate of decomposition were dramatically increased. Comparing the 4 hr weight loss to the cumulative weight loss after eight 30 min cycles showed that in each case the weight loss was more severe for cyclic heating than that observed during long term heating.

Fiberite 976 (Fig. 4 and Table 3) when heated for twelve cycles at 200°C showed a cumulative weight loss of 0.88% with 48% of the weight loss occurring in the first cycle. Increasing the temperature to 250°C showed a

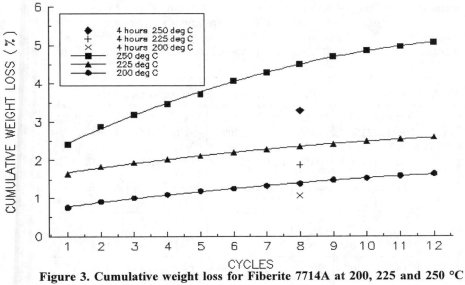

Figure 3. Cumulative weight loss for Fiberite 7714A at 200, 225 and 250 °C

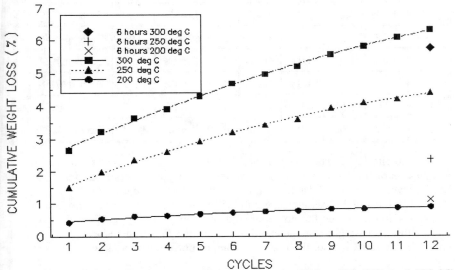

Figure 4. Cumulative weight loss for Fiberite 976 at 200, 250 and 300 °C

cumulative weight loss of 4.40% or a 400% increase. The rate of decomposition also showed a large increase. At 300°C, the weight loss increased to 6.32% and a slight increase in the decomposition rate was observed. Comparing the 6 hr weight loss to the cumulative weight loss after twelve cycles showed that at 250 and 300°C the cyclic decomposition was more severe, while at 200°C, the cyclic weight loss decreased slightly.

Table 3: Cumulative weight loss for twelve cycles at 30min/cycle

CY CLE	7714A 200°C WT. LOSS	7714A 225°C WT. LOSS	7714A 250°C WT. LOSS	976 200°C WT. LOSS	976 250°C WT. LOSS	976 300°C WT. LOSS	977-2 200°C WT. LOSS	977-2 250°C WT. LOSS	977-2 300°C WT. LOSS
1	0.75	1.65	2.42	0.42	1.50	2.66	0.34	1.58	3.46
2	0.90	1.83	2.88	0.54	1.99	3.24	0.41	2.01	4.15
3	1.00	1.93	3.18	0..61	2.35	3.64	0.46	2.28	4.70
4	1.09	2.02	3.46	0.63	2.60	3.93	0.49	2.48	5.16
5	1.19	2.11	3.72	0.68	2.94	4.31	0.54	2.87	5.65
6	1.26	2.20	4.07	0.72	3.20	4.69	0.57	2.97	6.08
7	1.33	2.28	4.29	0.75	3.43	4.97	0.60	3.08	6.49
8	1.39	2.35	4.51	0.77	3.59	5.20	0.62	3.18	6.86
9	1.48	2.42	4.71	0.82	3.95	5.58	0.68	3.30	7.24
10	1.53	2.49	4.86	0.83	4.11	5.82	0.75	3.38	7.58
11	1.59	2.55	4.97	0.86	4.21	6.09	0.77	3.45	7.81
12	1.69	2.60	5.08	0.88	4.40	6.32	0.78	3.52	8.25

Fiberite 977-2 when heated for 12 cycles at 200°C showed a cumulative weight loss of 0.78% with 44% of the weight loss resulting from the first heating cycle (Fig. 5 and Table 3). Increasing the temperature to 250°C resulted in the weight loss increasing to 3.52% or a 450% increase. The rate of decomposition increased but not as great as that observed in the 976 resin. At 300°C, the total weight loss increased to 8.25% or a 234% increase as compared to 250°C. Also, the rate of decomposition showed a large

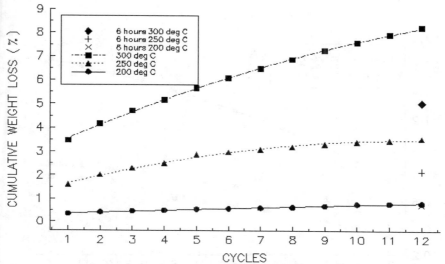

Figure 5. Cumulative weight loss for Fiberite 977-2 at 200, 250 and 300 °C

increase. In comparison to the 6 hr weight loss, at 250 and 300°C the cyclic weight loss was again much greater however, at 200°C, both results were almost identical.

CONCLUSION

Fiberite 976 and 977-2 (Fig. 6) both showed good thermal stability at 200°C based on both the continual and cyclic heating data. This is 23°C above the recommended service temperature for the former and 51°C above the latter. The 7714A resin system however, began to show significant decomposition at this temperature. Based on the 4 hr weight loss data, this material could probably sustain temperatures up to 150°C, which is 70°C above the recommended temperature.

Increasing the temperature to 250°C (Fig. 7) increased the material weight loss significantly for all three resins. Fiberite 977-2 showed the best stability, but from the data all three materials showed permanent decomposition.

At 300°C (Fig. 8) both 976 and 977-2 showed a substantial weight loss and significant decomposition. Fiberite 7714A was not tested at this temperature for obvious reasons.

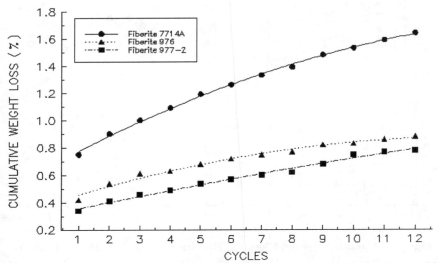

Figure 6. Cumulative weight loss for Fiberite 7714A, 976 and 977-2 at 200 °C

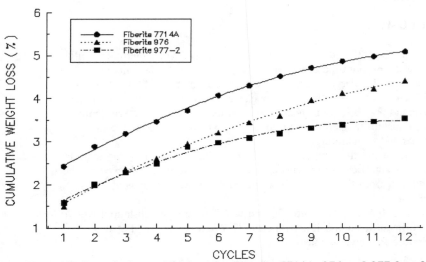

Figure 7. Cumulative weight loss for Fiberite 7714A, 976 and 977-2 at 250 °C

In summary, based on the data the onset of thermal decomposition begins between 150 and 200°C for Fiberite 7714A and 977-2, with decomposition beginning between 200 and 250°C for 977-2. Also, the effect of cyclic heating appeared to be more deleterious to the resin systems than continual heating, once the onset of decomposition begins.

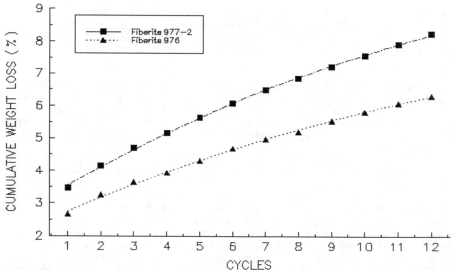

Figure 8. Cumulative weight loss for Fiberite 976 and 977-2 at 300 °C

REFERENCES

[1] Engineered Materials Handbook, Vol.1, Composites, American Society for Metals, Metals Park, Ohio, 1987.

[2] Wendlandt, W. W., Thermal Analysis, 3rd ed., John Wiley and Sons, New York, 1986.

[3] Turi, E. A., Thermal Characterization of PolymericMaterials, Academic Press Inc., Florida, 1981.

G. Mohammed Memon[1] and Brian H. Chollar[2]

Laboratory Simulation of Oxidative Aging of Asphalt

REFERENCE: Memon, G. M. and Chollar, B. H., "**Laboratory Simulation of Oxidative Aging of Asphalts**," Oxidative Behavior of Materials by Thermal Analytical Techniques, ASTM STP 1326, A. T. Riga and G. H. Patterson, Eds., American Society for Testing and Materials, 1997.

ABSTRACT: The Strategic Highway Research Program (SHRP) was funded by Congress in 1987-1993 to study the most pressing problems in the highway field. One of the major items studied in this program was the aging and changes of asphalt in pavements during construction of and during the life of the pavement.

The objective of this study was to determine the aging characteristics of asphalt mixes, which were mixed and conditioned at times corresponding to times experienced during field mixing. This objective was accomplished by measuring functional group changes of carbonyl and sulfoxide peak areas through infrared spectroscopy techniques of asphalts extracted from mixes subjected to various laboratory and climatic conditioning procedures. Correlations were attempted between these infrared functional group changes of asphalt samples from laboratory aged mixes and that from a pavement mix obtained through drum dryer plant production.

Ratios were taken of these unknown peak areas by dividing them by the area of a peak unchanged by oxidation, i.e., the methyl peak. These ratios were evaluated for asphalt mix samples that had undergone a wide range of treatments (oven heating for 0, 1, 2, 3, and 4 hours). It was observed through the carbonyl group changes of the asphalts that an oven aged period of 2 hours for asphalt mixes was needed to simulate the degree of aging occurring during plant mix production.

KEY WORDS: asphalt, oxidation, aging, mix, infrared, carbonyl group, sulfoxides.

---

1. Research Scientist, FHWA/SaLUT Inc., McLean, Va. 22101.

2. Research Chemist, FHWA, McLean, Va. 22101.

INTRODUCTION:

Asphalt is a by-product of the petroleum refining industry. Approximately 30 million tons of asphalt are being used in United States each year for road construction and repair. The oxidation of asphalt binder with age in the pavement causes hardening of the asphalt binder and shortens the life of the pavement. Interest has always been in the general aging, oxidation, and other changes occurring in asphalts as they are heated, mixed with aggregates, used in preparing pavements, and conditioned in the pavement under years of constant temperature and environmental fluctuations. The FHWA has been studying aspects of these changes through many projects throughout the years. One recent study involved obtaining the proper laboratory heating conditions of asphalt-aggregate mixes to simulate the changes occurring with asphalt and the mix in preparing a pavement. It involved the measurement of the physical properties of asphalt-aggregate mixes prepared in the laboratory after oven heating for 0, 1, 2, and 4 hours at 135°C. These data were compared with the physical properties of the mix obtained from freshly prepared and compacted pavement using the same asphalt and aggregate. The chemical properties were also measured on the extracted asphalts from mixes in this study to see the changes that occurred in the asphalts upon heating and interaction of aggregates and to show the laboratory conditioning that is most similar to that occurring in the preparation of the pavement. The infrared study presented here is part of this study.

BACKGROUND:

The Strategic Highway Research Program (SHRP) was an independent program funded by Congress in 1987-1993 to study the most pressing problems in the highway field and to obtain implementable products for States and the highway community in that short time period. One of the major items studied in this program was the aging and changes of asphalt in pavements during construction of and during the life of the pavement. Some of the major products to come forth from SHRP were specifications for the use of asphalt binders to be used in pavements and for mix designs using aggregates and asphalts for the construction of pavements.

One of the test procedures in the SHRP mix specification calls for conditioning the desired mix of asphalt and aggregate in an oven at 135°C for 4 hours[2]. This conditioning simulates the conditioning that the mix would be subjected to before it is laid down on a pavement and compacted at the job site. FHWA personnel questioned the time of oven heating of mixes as being too harsh a condition for accurate simulation of the field mix. Since FHWA was constructing pavements on a field site near their research laboratories for accelerated load studies for pavement fatigue and permanent deformation, FHWA personnel began a laboratory study to see what time period of oven conditioning of mixes of one aggregate and three asphalts used in this pavement best simulates the mix that is prepared by a hot mix plant and is freshly laid down on a pavement[1]. An AC-5, AC-20, and a Styrelf (a polymer modified asphalt) were used in this oven aging study. Mixes representative of a typical surface pavement layer used in this study were made in the laboratory in triplicate using these asphalts and the aggregate and heated in a laboratory oven for 0, 1, 2, and 4 hours. In addition, A mix typical of a base layer used in this pavement load study was made using the same AC-5 asphalt and aggregate and subjected to the same laboratory heating conditions. A total of 48 laboratory mixes (3+1 asphalts x 4 conditioning times x 3 replicates) were prepared and conditioned[1].

These 48 mixes were compacted into cores using laboratory Marshal compactors under conditions known to represent field compacting conditions. Field cores were also taken from the newly constructed accelerated load facility pavements. Cores were taken from these pavements 2 weeks old that contained the surface gradated mixes of the AC-20, the Styrelf asphalt, and the surface and base gradated mixes of the AC-5 asphalt.

A mix related battery of tests was conducted with the laboratory and field cores. As part of these tests, the asphalts were extracted from the cores to obtain the percentage of asphalt in the cores. The extracted asphalts were saved for further physical characterization using the SHRP binder specification equipment. The physical data of the asphalt and the mixes from these studies showed that the laboratory conditioning of mixes for 4 hours, as prescribed in the SHRP procedures, were too long a time period to simulate the field construction conditions. At that point, FHWA decided to look at the chemical changes of the asphalts from this laboratory conditioning and the comparison to that of field conditioning. The study presented here is part of this chemical study for asphalt laboratory and field comparison and changes.

PURPOSES OF THE STUDY:

The purposes of this study are to 1) determine the aging characteristics and other changes of asphalts occurring when mixes containing these asphalts are subjected to oven conditioning for various periods of time, 2) determine the time period of mix laboratory conditioning necessary to simulate the conditions field mixes are subjected to before and during pavement preparation. These objectives will be accomplished by measuring functional group changes through infrared spectroscopy techniques of asphalts extracted from mixes subjected to various laboratory and field conditioning procedures.

HYPOTHESIS:

There is significant difference in the peak areas of the functional groups, C=O and S=O, for asphalts in mixes conditioned by heat for up to 4 hours.

VARIABLES:

The independent variables are time of oven conditioning of mixes for surface AC-5 asphalt and the field conditioning procedure for mixes. The dependent variable is the peak area of the various functional groups in the asphalts.

SCOPE:

The validation of the suggested Strategic Highway Research Program (SHRP) laboratory mix procedure for simulating/predicting the plant mixing and pavement lay down procedure is being conducted in this study by characterizing and comparing infra red spectra of binder extracts from the field pavement cores and laboratory mixes.

The thermal oxidative technique used was the conditioning of laboratory mixes by heating samples in a laboratory oven for 0, 1, 2, and 4 hours. The interaction of asphalt with aggregate under these laboratory conditions were being studied to try to simulate the natural oxidative conditioning of asphalts in pavements occurring in the field. Infrared spectra of samples of asphalts from AC-5 (surface) oven conditioned mixes(12) and from replicates of the field conditioned mixes (4) were obtained. Areas of significant peaks representing asphalt functional groups were obtained by using an infrared software program (Macro/OMNIC) to obtain I. R. Spectra.The functional group peaks that significantly change with mix laboratory conditioning were identified and used in the comparisons.

Using these selected peak areas, the type of functional group changes upon mix laboratory conditioning was interpreted along with reasonable explanations as to why these changes occurred.

Field mix asphalt peak areas were compared with laboratory mix asphalt peak areas to obtain the period of mix laboratory conditioning time that most simulates the field mix conditioning procedures.

These results were compared to those physical results obtained by FHWA personnel both with these asphalts and mixes. SHRP results pertaining to this conditioning procedure were also examined.

PROCEDURES:

Asphalt mixes were prepared in the laboratory using an AC-5 and an aggrigate and conditioned in oven according to AASHTO procedure P2-94[2].
The asphalts were recovered using an Abson recovery method from the FHWA laboratory mix conditioning study for the accelerated load facility pavements were used in this study.

A Nicolet Fourier Transform Infrared Spectrometer (FTIR), Model 730, was used to obtain the infrared (IR) spectra of the different asphalts used in this study.   Sixty scans of background as well as sixty scans of each IR sample were taken. The data from these were averaged to obtain a single spectrum of the sample. A thin film of hot asphalt was drawn on a sodium chloride cell. The infrared spectrum was obtained for the asphalt sample for the frequency range of 400-4000 cm-1. The thickness of the asphalt film was adjusted until the transmittance spectrum of the hydrocarbon peak at 2875 cm-1 was between 12-18 percent (bottom of the spectrum) and 80-90 percent (top of the spectrum). Two spectra of each asphalt replicate (R1, 2, 3) for a set of heating conditions were obtained. The two spectra for each asphalt replicate were analyzed by a Macro written in the OMNIC IR software to show differences in functional groups in asphalt samples. The spectra selected were then further processed under a "macro process" protocol (written in our laboratory, which is a very successful way of producing repeatable peak area's) which generated the base line correction and the peak area's for figures 2, 3, 4, and 5. In this analysis the five peak areas and peak to peak ratios were calculated in each spectrum as are shown in figure 1. The ratios and peak areas were taken after normalization with the CH_2, CH_3 signal (1327-1526 cm^{-1}). The ratios and peak areas for each spectra were compared to show more precisely the differences in functional groups in the spectra.

RESULTS AND DISCUSSION:

The results discussed in this work were obtained using only those areas which have some significant impact on aging such as the carbonyl group (C=O) and the sulfoxide group (S=O). It is already reported in the literature that the oxidation of asphalt and resulting carbonyl compound formation in the range of 1640-1760 cm^{-1} changes the physical properties of the asphalt binder[4-6] and is the major cause of several types of pavement failure.

Figure 2 shows the peak areas of the total C=O group {the total C=O (1548-1808 cm^{-1})is the original carbonyl group and the newly developed C=O (1642-1742 cm^{-1}) group formed on aging} versus the time period of laboratory mix aging. The total C=O peak area shows some inconsistency with aging time. However, when total C=O area of the field sample was correlated with the laboratory aged samples, the field mix conditioning was simulated by the laboratory mix conditioning of a little more then 3 hours.

Figure 3 shows the peak area of the newly developed C=O (after aging) (1642-1742 cm^{-1}) versus time of laboratory aging. It was observed in the field as well as laboratory samples that a new C=O group was formed in the recovered asphalts. It is our contention this new C=O signal is responsible for the field oxidation. The new carbonyl peak area of the field asphalt sample is most similar to that of the 2 hrs laboratory conditioned asphalts. Secondly, the laboratory mix aging shows good consistency in the peak areas with the r^2 value of 0.967. According to the standard practice for short and long term aging of hot mix asphalt (AASHTO P2-94)[2], a mix consisting of aggregate and asphalt binder is aged in a forced draft oven for 4 hours at 135^{0}C. However, the new carbonyl peak areas for asphalts obtained through laboratory mix aging shows that 2 hours of aging can simulate the field conditioned asphalt samples. Stuart et al[1] suggested that an oven aging period of 2 hours was needed to simulate the average degree of aging that occurred during plant production. They used the G*/Sin parameter for comparisons with the mix testing protocol for prediction purposes.

Figure 4 shows the comparison of old C=O (the original C=O group of asphalt) peak area (1548-1642 cm^{-1}) versus the time of aging. From this peak area simulation, the field sample shows that 1.2 hours laboratory mix aging is equal to the field conditioning of asphalt samples. The regression analysis of the laboratory mix aging samples shows an r^2 value of 0.814.

Figure 5 shows for the comparison of S=O group (985-1086 cm^{-1})(another asphalt functional group usually showing changes upon asphalt aging) versus time of aging. The field conditioning of asphalt samples shows a S=O peak area similar to 2.7 hours laboratory mix aging. However regression analysis shows an r^2 value of 0.37, a poor correlation. Mill[3] has found that the amount of sulfoxides is reduced upon aging due to the formation of C=O. In other words, the sulfoxide is helping in the formation of C=O. That may be the reason that this is not a good correlation. Further, the S=O infrared signal did not give a consistent trend of increased oxidation in peak areas; it first increased and then decreased over the time period both in field as well as laboratory mixes. However

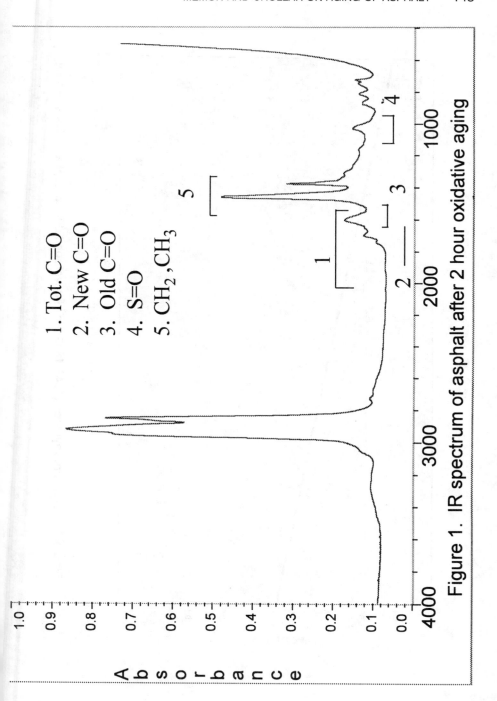

Figure 1. IR spectrum of asphalt after 2 hour oxidative aging

1. Tot. C=O
2. New C=O
3. Old C=O
4. S=O
5. CH_2, CH_3

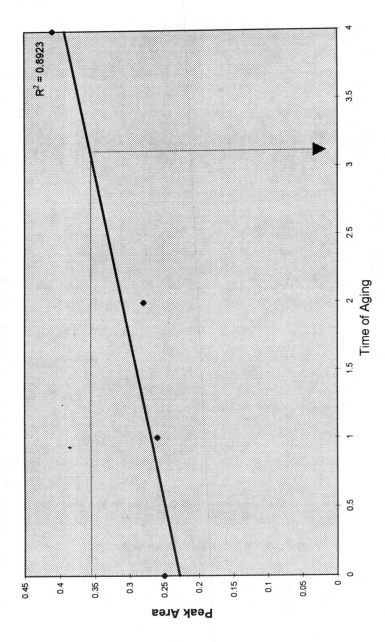

Figure 2: Total carbonyl area (1548-1808 cm^{-1}) developed during asphalt mix oven aging

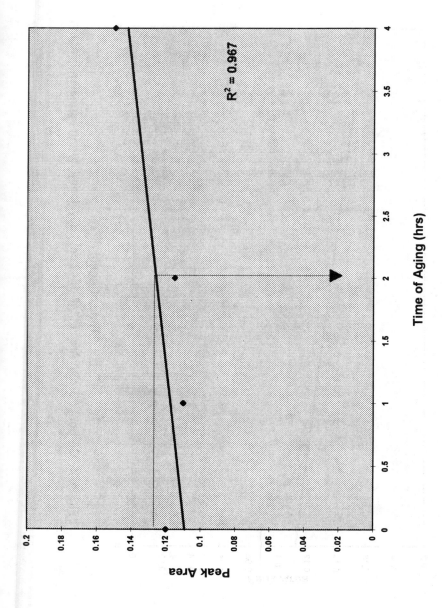

Figure 3. Carbonyl group area (1642-1742 cm^{-1}) developed during asphalt mix oven aging

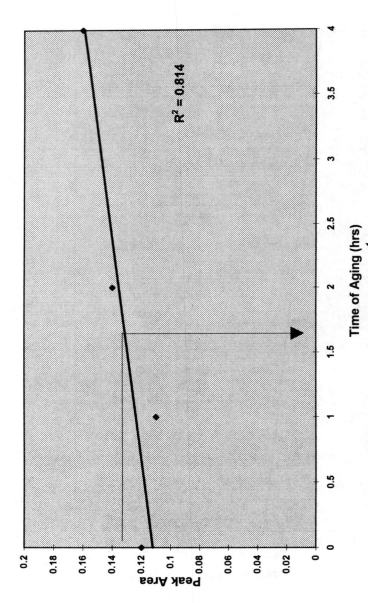

Figure 4. Carbonyl Group Area (1548-1642 cm^{-1}) developed during asphalt mix aging

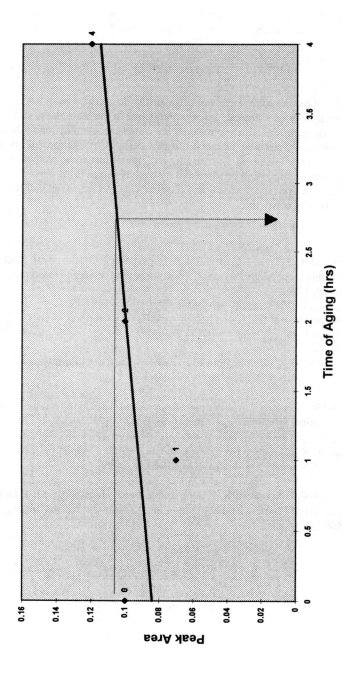

Fidure 5. Sulfoxide Group (985-1086 cm^{-1}) Developed During Asphalt Mix Oven Aging

the C=O infrared signal showed a consistent increase over the time period. Thus it was more reliable to predict the oxidative behavior of asphalts from C=O signal then S=O signals.

CONCLUSIONS:

1. Carbonyl groups and sulfoxide groups in asphalts do increase as mixes containing these asphalts are conditioned, either with heating in the laboratory, during drum dryer or pug mill operations, or the pavement lay-down and compaction in the field.

2. Data from infrared carbonyl peak areas of new carbonyl peaks formed upon asphalt oxidation show that field drum dryer or pug mill operations and pavement lay-down and compaction procedures can be simulated by heating mixes in an oven at 135°C for 2 hours. Both the studies of the chemical and physical changes in asphalt and the physical changes of the mix verify this.

REFERENCES:

1.   K. D. Stuart and R. P. Izzo, " Correlation of Superpave™ G*/Sin with Rutting Susceptibility from Laboratory Mixture Tests", Transportation Research Board preprint # 950917, January 22-28, 1995.

2.   American Association of the State Highway and Transportation Officials (AASHTO), P2-94, pp15-16.

3.   T. Mill, D. S. Tse, C. C. D. Yao, and E. Canavesi, "Oxidation Pathways in Asphalt, pp 1367-1375, 204th ACS Meeting, Washington D. C., 1992.

4.   D. Y. Lee and R. J. Huang, "Weathering of Asphalt as Characterized by Infrared Multiple Internal Reflection Spectra," Anal. Chem., **46**, 2242, 1973.

5.   K. L. Martin, R. R. Davison, C. J. Glover, and J. A. Bullin,"Asphalt Aging in Texas Roads and Test Sections," Trans. Res. Rec., **1269**, 9, 1990.

6.   J. C. Petersen, J. F. Branthaver, R. E. Robertson, P. M. Harnsberger, J. J. Duval, and E. E. Ensley,"Effects of Physicochemical Factors on Asphalt Oxidation Kinetics," Trans. Res. Rec., **1391**, 1, 1993.

Instrumental Techniques—
Oxidation Induction Processes

Alan T. Riga[1] and Gerald H. Patterson[1]

DEVELOPMENT OF A STANDARD TEST METHOD FOR DETERMINING OXIDATIVE INDUCTION TIME OF HYDROCARBONS BY DIFFERENTIAL SCANNING CALORIMETRY AND PRESSURE DIFFERENTIAL SCANNING

REFERENCE: Riga, A. T., and Patterson, G. H., "**Development of a Standard Test Method for Determining Oxidative Induction Time of Hydrocarbons by Differential Scanning Calorimetry and Pressure Differential Scanning Calorimetry,**" *Oxidative Behavior of Materials by Thermal Analytical Techniques, ASTM STP 1326,* A. T. Riga and G. H. Patterson, Eds., American Society for Testing and Materials, 1997.

ABSTRACT: Test methods under the auspices of ASTM Committee E37 on Thermal Measurements are being developed for determining the oxidative induction time (OIT) of hydrocarbons by differential scanning calorimetry (DSC) and pressure DSC. This test method,ASTM E1858, is applicable to hydrocarbons, for example, polyolefins and motor oils that oxidize exothermically in their analyzed form. The DSC method is used at ambient pressure, one atmosphere oxygen. The pressure DSC (PDSC) protocol is used at high pressure, 3.5 MPa or 500 psig , oxygen. The test specimen and reference aluminum pan are heated to a specific temperature, DSC at 195°C and PDSC at 175°C in an oxygen environment. Heat flow out of the specimen is monitored at the isothermal temperature until the oxidative reaction is manifested by heat evolution on the thermal curve.

Statistically designed experiments aided in establishing the factors affecting the OIT. Sample pan type, pressure, temperature, and oxygen flow rate were observed as significant experimental parameters. The OIT is a relative measure of oxidative stability at the test temperature. A precision and bias statement has been developed by ASTM Committee E37 and is reported based on a recent national round robin.

KEYWORDS: oxidation, oxidation induction time(OIT), Differential Scanning Calorimetry (DSC), Pressure Differential Scanning Calorimetry (PDSC), hydrocarbons, polyolefins, oxidative stability, isothermal temperature, standard deviation and precision

There is an ongoing need in academic or industrial research for standard test methods to determine the oxidative properties of materials. ASTM committees have dealt with this need for specific oxidation requirements. There is a standard test method for the OIT of

1 Senior research chemist and technology manager, respectively, Lubrizol Corporation, 29400 Lakeland Blvd., Wickliffe, OH 44092.

polyolefins by DSC [ASTM Test Method for Oxidative Induction Time of Polyolefins by Differential Scanning Calorimetry (D3895)].There is also a method for lubricating grease oxidation by PDSC [ASTM Test Method for Oxidation Induction Time of Lubricating Greases by Pressure Differential Scanning Calorimetry (D5483)]. Polyethylene plastic pipe and fitting materials can be characterized by a DSC oxidative stability method [ASTM Specification for Polyethylene Plastics Pipe and Fittings Material (D3350)]. DSC is also used in a standard test method [ASTM Test Method for Physical and Environmental Performance Properties of Insulations and Jackets for Telecommunications Wire and Cable (D4565)] for physical environmental performance properties of insulation and jackets for telecommunications wire and cable. Typical uses of an OIT test method include oxidative stability of edible oils and fats (oxidative rancidity), lubricants, greases and polyolefins [1-6].

The scope of this study is to define test methods to determine the oxidative properties of hydrocarbons which exotherm in their analyzed form by DSC or PDSC [ASTM Terminology Relating to Thermal Analysis (E473)]. The test specimen in an aluminum pan is heated to a specific isothermal test temperature in an oxygen environment. Heat flow out of the specimen is monitored at an isothermal temperature until the oxidative reaction is manifested by heat evolution on the thermal curve.

The OIT, a relative measure of oxidative stability at the test temperature, is determined from data recorded during the isothermal DSC test, Figure 1. The presence or effectiveness of antioxidants may be determined by this method. The OIT values obtained may be compared from one hydrocarbon to another or to a reference material to obtain relative oxidative stability information.

Experimental Design
Some of the factors affecting the OIT test are: isothermal temperature, pressure, sample size, gas flow rate, and pan surface metallurgy [7,8]. Others considered but not studied are: catalyst, gas type (as air or oxygen), and heating rate (40°C/min) to the isothermal temperature.

The pressure effect at 195°C on the OIT is viewed in Fig. 2. High pressure reduces the OIT from 20 to 7.5 min. The temperature effect at 3.5-MPa oxygen on the OIT is seen in Fig. 3. Higher temperatures reduce the OIT from 30 to 7.5 min in this example.

An experimentally designed study used the following conditions: **sample**: diluted motor oil (ASTM reference B), **sample size**: 1.00 and 3.00 mg, **gas flow rate**: 25 and 100 mL/min, **pressure**: 80 and 500 MPa, **isothermal temperature**: 175 and 195°C, and **pan type**: aluminum or anodized aluminum (aluminum oxide coated pans).

A two to the third factorial design based on the oxidation induction time and the variables of pan type, flow rate and pressure can be seen in Fig. 4. Dot plots of the factor

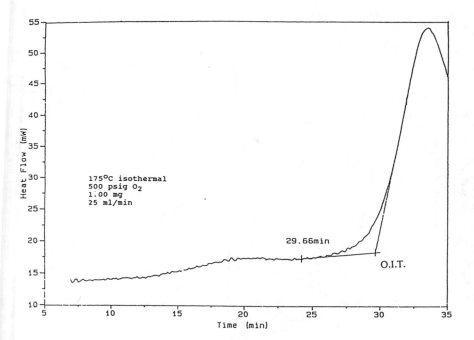

Figure 1. DSC Oxidation Induction Time,
extrapolated onset time.

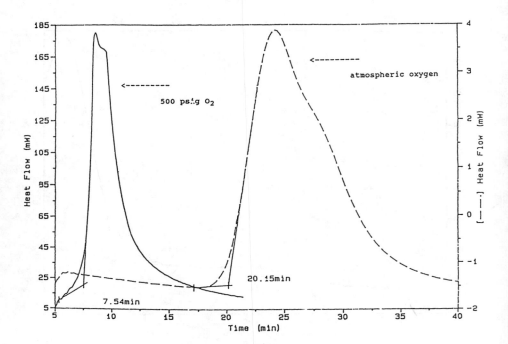

Figure 2. The Pressure Effect at 195°C on the
Oxidation Induction Time

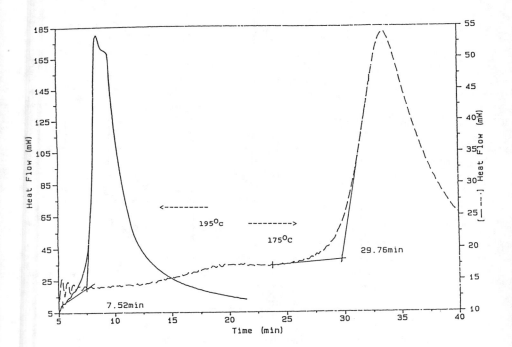

Figure 3. The Temperature Effect at 3.5 Mpa oxygen on the
Oxidation Induction Time

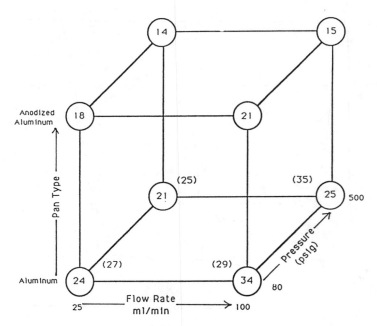

Figure 4. Two to the Third Factorial Design:
Oxidation Induction Time by DSC,
OIT values enclosed in circles
(repeated values in parentheses)

Experiment 1:   2³ Factorial Design -- 8 Runs

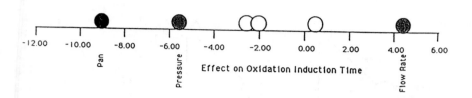

Experiment 1A: 2³ Factorial Design -- 8 Runs with 4 Repeats

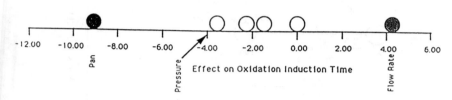

Figure 5. Dot Plots of the Factor Effects in a Two to the
Third Factorial Design.

effects in this design are recorded in Fig. 5. Pan type and pressure had a negative effect
on the OIT, while the flow rate had a slight positive effect on the OIT.

A two to the fourth factorial design was also used to evaluate the effects of sample mass,
flow rate, flow rate direction (reverse flow), and pan type, see Figs. 6 and 7. The sample
size and flow rate direction had a slight, 4.0 min, positive effect on the OIT, while the
pan type definitively decreased the OIT by 10 to 12 mins. Interactive variables for both
factorial designs were considered and found to be insignificant.
The experimental design set some of the parameters for this ASTM procedure (see Figs.
5 and 7): sample size of 3.00 mg and the flat aluminum pan, for sample and reference,
oxygen pressure of one atmosphere at the preselected isothermal temperature of 175°C,
an oxygen pressure of 3.5 -MPa at the also preselected isothermal temperature of 195°C,
and the gas flow direction was recommended in the "reverse" flow mode.

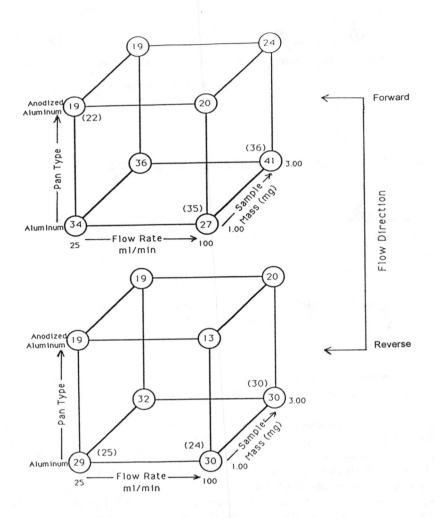

Figure 6. Two to the Fourth Factorial Design:
Oxidation Induction Time by DSC,
OIT values enclosed in circles
(repeated values in parentheses)

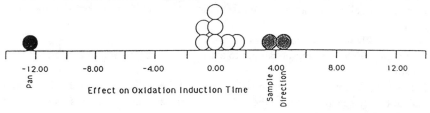

Figure 7. Dot Plots for the Two to the Fourth Factorial
Design: Oxidation Induction Time by DSC

Experimental Procedures [1] (Task Group E37.01.10)
A differential scanning calorimeter or pressure differential scanning calorimeter is used that is capable of providing a heating rate of 40°C/min and automatically recording the power difference or the differential heat flow input between the sample and reference material. The calorimeter should have the capability of measuring the heat flow of at least 5 mW. Isothermal temperatures must be held within ± 0.4°C of set temperature.

The time base is measured to within ± 0.1 min. The capability to record the first derivative of the heat flow curve will be helpful in cases in which the baseline is not constant. High pressure oxygen is set at 3.5-MPa oxygen (500 psig). The oxygen flow rate is 50 mL/min and is accurate to within ± 5%. The sample mass is nominally 3 mg. Specimen capsules or holders are the aluminum sample pans and should be inert to the sample and the oxidizing gas. The pans shall be clean, dry, and flat.

The materials used in this study were oxygen, extra dry, of not less than 99.5% purity by volume. The calibrants, indium and tin, are 99.9% purity by mass.

If the sample is a liquid or powder, mix thoroughly before sampling. In the absence of specific sampling information, samples are to be analyzed as received. If some heat or mechanical treatment is applied to the sample before analysis, this treatment should be in nitrogen and noted in the report. If some heat treatment is used before oxidative testing, then record any mass loss as result of the treatment.

Calibrate the temperature output of the calorimeter using ASTM Practice for Temperature Calibration of Differential Scanning Calorimeters and Differential Thermal Analyzers(E967), except that a heating rate of 1°C/min is used to approach the isothermal conditions of this test. Use indium and tin as calibration materials to bracket the temperature used in the PDSC (175°C) and DSC (195°C) tests. Calibration shall be performed under ambient pressure conditions. The melting temperatures observed in the calorimeter calibration are obtained from the extrapolated onset temperatures.

DSC and PDSC Test Procedures

1. Weigh 3.00 to 3.30 mg of sample to a precision of ± 0.01 mg in a clean specimen capsule.
2. Place the uncovered specimen containing pan in the sample position of the calorimeter and an empty specimen pan in the reference position. Center the pans on the sensors.
3. Adjust flow rate of oxygen gas to 50 mL/min.
4. Set the instrument sensitivity as required to retain the oxidation exotherm on the recorded range. A sensitivity of 2 W/g full scale is typically acceptable.
5 Purge the specimen area for 3 to 5 mins to ensure exchange of air with oxygen at atmospheric pressure. Check the flow rate at elevated pressure and readjust to 50 mL/min, if required.
6. Program heat at 40°C/min from ambient temperature to the isothermal temperature, 175 or 195°C. Wait until the temperature reaches isothermal conditions and then record the thermal curve.
7. The isothermal temperature should be maintained at ±0.4°C.
8. Zero time is recorded at the initiation of the OIT measurement.
9. The OIT is the total time from the start of the experiment at room temperature in oxygen to the extrapolated onset time of the exothermic process.
10. Test Methods
Method A. When using the **DSC Test Method**, maintain flow rate of 50 mL/min of oxygen and an isothermal temperature of **195°C.**
Method B. When using the **PDSC Test Method,** pressurize slowly, adjust and maintain pressure of oxygen at 3.5 MPa and a flow rate of 50 mL/min and an isothermal temperature of **175°C.**
11. Continue isothermal operation until the peak of the oxidation exotherm is observed or until an inflection point is observed and the total displacement from the initial baseline exceeds 3 mW (1 W/g).
12. Determine the OIT. Extend the recorded baseline at the isothermal temperature beyond the oxidation reaction exotherm see Fig. 1. Extrapolate the slope of the oxidation exotherm from the inflection point on the curve to extended baseline. Determine the time at the intersection; it is the OIT.
13. Report the following values: identification of the sample, method used A (DSC) or B (PDSC), oxygen pressure in MPa and flow rates in mL/min, Isothermal temperature, OIT in minutes, specimen mass and description of apparatus.

Results and Discussion

For some particularly stable materials, the OIT may be quite long, for example > 120 mins, at the specified elevated temperatures of the experiment. Under these circumstances, the OIT may be reduced by increasing the isothermal temperature or increasing the pressure of the oxygen purge gas or both. Reactions that proceed too rapidly, with a short OIT, may be extended by decreasing the test temperature or reducing the partial pressure of oxygen or both. By admixing oxygen gas with a suitable diluent, for example, nitrogen, the OIT will be increased.

For some systems, the use of copper pans to catalyze oxidation will reduce the OIT for a given temperature. The results, however, will not correlate with noncatalyzed tests.

In certain cases when the sample under study is of high volatility, for example, a low molecular weight hydrocarbon, either the use of pressures in excess of one atmosphere or lower temperatures may be required.

An interlaboratory test [ILT] program based on a 1995-1996 Interlaboratory study was carried out to determine the repeatability and reproducibility of the test methods, ASTM Standard Practice for Conducting an Interlaboratory Study to Determine the Precision of a Test Method (E691). The ILT was divided into four parts: a lubricating oil sample C and a polyethylene film sample D which were examined in duplicate by the above cited protocol. A standard DSC at 10-KPa oxygen pressure and a pressure DSC at 3.5-MPa oxygen pressure were used in the ILT. A total of eight laboratories performed the DSC test and seven laboratories the PDSC test.

The statistical analysis for the oil and polymer at two oxygen pressures is summarized in Table 1.

TABLE 1 - *ASTM Committee E37.01.10 Interlaboratory Test*

Reference Material	Oxygen Pressure	Temperature (C)	OIT mean (min)	Sr (min)	RSr (%)	SR (min)	RSR (%)
Lube Oil C	10 KPa	195	62.2	3.4	5.5	8.2	13.1
Lube Oil C	3.5 MPa	175	42.0	1.1	2.6	10.3	24.6
Polyethylene D	10 KPa	195	29.0	1.3	4.3	4.5	15.4
Polyethylene D	3.5 MPa	175	25.6	1.7	6.7	1.9	7.4

OIT = mean Oxidation Induction Time
Sr = repeatability standard deviation
Rsr = relative repeatability standard deviation (Sr/mean OIT X100)
SR = reproducibility standard deviation
RSR = relative reproducibility standard deviation (SR/mean OITx100)

The lube oil sample C , characterized with this standard method at 195°C and 10-KPa oxygen pressure, had a mean OIT value of 62.2 mins. The repeatability standard deviation was 3.4 mins (5.5%) and a reproducibility standard deviation of 8.2 mins (13.2%).

The lube oil sample C, tested with this standard method at 175°C and 3.5-MPa oxygen, 500 psig, oxygen pressure, had a mean OIT value of 42.0 mins. The repeatability standard deviation was 1.1 mins (2.6%) and a reproducibility standard deviation of 10.3 mins (25%).

The polyethylene sample, D, characterized with this standard at 195°C and 10 Kpa oxygen pressure, had a mean OIT value of 29.0 mins. The repeatability standard deviation was 1.3 mins (4.5%) and a reproducibility standard deviation of 4.5 mins (16%).

Polyethylene sample D was also tested under the auspices of ASTM Committee D20 Plastics by PDSC, but under different conditions [5]. The latter PDSC test was run at 200°C and 10 kPa oxygen pressure. The OIT was 31.4 mins with a repeatability standard deviation was 1.6 mins (5.1%) and a reproducibility standard deviation of 3.1 mins (9.8%). The results of the studies by ASTM Committees E37 and D20 are in agreement, since there is no statistical significance of the mean values at the 95% confidence level.

Polyethylene sample D was tested with the ASTM Committee E37 protocol at 175°C and 3.5-MPa oxygen, 500 psig, oxygen pressure had a mean OIT value of 25.6 mins. The repeatability standard deviation was 1.7 mins (6.7%) and a reproducibility standard deviation of 1.9 mins (7.4%).

Conclusions

A standard test method has been developed to determine the oxidative behavior of hydrocarbon oils or polymers by measuring the oxidation induction time at selected isothermal temperatures and oxygen pressures. This test method is repeatable and reproducible.

Acknowledgments
The authors are pleased to acknowledge the assistance of Roger Blaine, TA Instruments for the Interlaboratory Summary, the companies who participated in the OIT study, the Lubrizol Corporation, the Thermal Analysis Laboratory and Statistical Services, Amoco Fabrics, Albermarle, TA Instruments, Mettler-Toledo Instruments, Raychem, Petro-Canada, and the Perkin Elmer Corporation.

References

[1] Blaine, R. L. , *NLGI Spokesman*, June, 1976, p. 94.
[2] Noel, F. , Journal of the Institute of Petroleum London, Vol. 57, No. 558, 1971, p. 354.
[3] Hsu, S. , Cummings, A. , and Clark, D. , *Society of Automotive Engineering*, Technical Paper 821252, 1980.
[4] Zalogina, K. , *Chemical Technology Fuels Oils*, Vol.18, 1982, p. 425.
[5] Zeman, A. , Stuwe, R. , and Koch, K. , *Thermochimica Acta*, Vol. 80, 1984, p. 1.
[6] Walker, J. , and Tsang, W. , *Society of Automotive Engineering*, Technical Paper 801383, 1980.
[7] Patterson, G. H. , and Riga, A. T. , "Factors Affecting Oxidation Properties in Differential Scanning Calorimetric Studies," *Thermochimica Acta*, Vol. 226, 1993, pp. 201-210.
[8] Stricklin, P. L. , Patterson, G. H. , and Riga, A. T. , " The Development of a Standard Method for Determining Oxidation Induction Times of Hydrocarbon Liquids by Pressure Differential Scanning Calorimetry," *Thermochimica Acta*, Vol. 243,1994, pp. 201-208.

R. Bruce Cassel,[1] Andrew W. Salamon, [1] Gregory Curran, [1] and Alan T. Riga[2]

COMPARATIVE TECHNIQUES FOR OXIDATIVE INDUCTION TIME (OIT) TESTING

REFERENCE: Cassel, R. B., Salamon, A. W., Curran, G., and Riga, A. T., "Comparative Techniques for Oxidative Induction Time (OIT) Testing," Oxidative Behavior of Materials by Thermal Analytical Techniques, ASTM STP 1326, A. T. Riga and G. H. Patterson, Eds., American Society for Testing and Materials, 1997.

Abstract: The oxidative induction time (OIT) test has been performed using a range of sample materials, instruments and conditions. Using the method of the ASTM E37.01.10 Task Group Interlaboratory Study as a basis, test samples of motor oil and high density polyethylene are analyzed using several analytical options. Instrumentation includes power compensation differential scanning calorimeter (DSC), pressurized cell power compensation DSC and heat flux DSC. The use of vented capsules as an alternative to the standard open capsules is also investigated.

The results show in the case of high density polyethylene (HDPE) (the primary sample studied for this contribution) that statistically similar results were obtained using power compensation DSC and heat flux DSC, and using three types of sample pans, one of which was a side-vented autosampler-compatible pan.

Keywords: Oxidative Induction Time, (OIT), differential scanning calorimeter, DSC, Power compensation DSC

Background

The oxidative induction time (OIT) test has been used in thermal analysis on hydrocarbon materials for more than 20 years [1]. The chemistry and physics of this thermal titration procedure are well known [2]. There have been numerous standard tests methods developed to be able to assay materials reliably with respect to their resistance to oxidative degradation [3]. When one of these tests is performed on a given piece of equipment under carefully controlled conditions, the results can be quite reproducible. However, often the results on a particular sample analyzed in two different laboratories are dramatically different. Several interlaboratory tests have been undertaken to establish

[1]Senior Scientists, The Perkin-Elmer Corporation, 50 Danbury Rd., Wilton CT 06897

[2]Senior Scientist, The Lubrizol Corporation, Wickliffe, OH 44092

the conditions under which results can be compared between different types of instrumentation.

In the development of standard tests, it is important that the method be applicable to a wide range of instrumentation. However, there are some considerations and terminology which are specific to power compensation type differential scanning calorimetry. These considerations are almost never covered in standard test methods because they are only applicable to one vender's instruments. This is unfortunate since it has no doubt resulted in a broader spread of test results than would have been obtained if the differences between the two analytical approaches were better understood. Since this ASTM publication will likely become the definitive reference for the OIT, test it does seem appropriate to clarify a few of these issues in this paper. Of course, for more detailed information on running OIT on any particular piece of instrumentation, it is advisable to contact the manufacturer for detailed specific advise.

Power Compensation and Heat Flux -type DSCs

Figure 1 illustrates the two types of DSC cell and their relevant nomenclature. In the case of power compensation DSC, the encapsulated specimen sits in its own "sample holder" which consists of a platinum calorimeter having its own heater, platinum resistance thermometer sensor. The weight of the calorimeter is around 1g. The furnace, sensor and sample holder floor are intimately coupled and maintained at nearly the same temperature using a two-decade feedback loop. As a result of the tight temperature control, a small sample specimen is within 0.1 °C of the final isothermal temperature within 1 min from the time that the heater starts heating from the load temperature, and within 30 sec of the time of the programmed start of the isotherm. (See Figure 2.) For an appropriately calibrated system, there will be less than 0.1°C of temperature overshoot; once the temperature has arrived it will hold constant within 0.1°C throughout the OIT experiment. This is an inherent feature of power compensation DSC, [4] and it is the same with the power compensation pressure DSC accessory.

In the case of heat flux DSCs the furnace encloses both the sample and reference capsule, hence, it is larger with a large thermal mass. In the case of pressure DSC, it may be especially massive. It is also less closely coupled to the sample. Older instruments may take several minutes to reach the isothermal temperature. They may also exhibit substantial temperature overshoot. Because of this problem, previous OIT methods have called for equilibration at the target temperature in nitrogen before switching to oxygen. This may result in evaporation of antioxidant and other volatiles before starting the antioxidant titration. Furthermore, there is a period during switchover when the partial pressure of oxygen is changing and the reaction is starting but at an indefinite rate, depending on the displacement rate of oxygen directly over the specimen. These problems are minimized with the ASTM Committee E37 OIT test (E1858) because the entire test is performed with an oxygen purge. Still, there is the problem of uncertain reaction start during thermal equilibration. Fortunately, most modern heat flux DSCs equilibrate sufficiently rapidly that this may not be a problem.

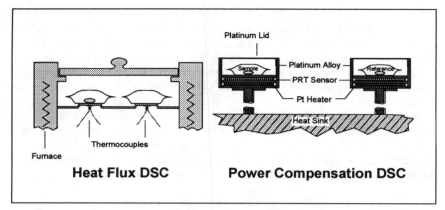

Fig. 1 - Heat Flux DSC Cell and Power Compensation DSC Sample Holders

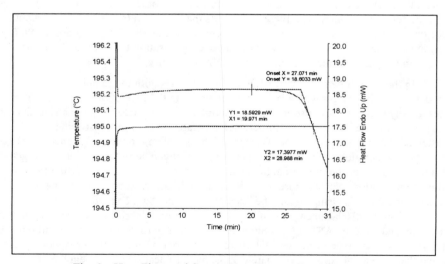

Fig. 2 - Heat Flow and Sensor Temperature Data from a
Power Compensation DSC: OIT of HDPE

Experimental Method

The method was developed for ASTM Committee E37 using the best features of other OIT methods and the considerable experience of Alan Riga and others on the E-37.01.10 subcommittee. The details of the method can be found in the final ASTM E1858 document and not all the details are repeated here. Certain problem areas are discussed below, especially where there are issues involving differences between heat flux and power compensation instrumentation.

Purge Conditions

The purge rate as recommended was 25 cm3/min measured at the purge gas outlet. For testing at elevated pressures the purge rate is measured at atmospheric pressure. The purge gas is purified oxygen throughout the test. This eliminates the problems of elevated temperature switchover discussed above. Time (5 min) was allowed at the loading temperature for residual nitrogen in the loading atmosphere to be replaced by oxygen. This procedure is especially important for using vented sample pans which are necessary for using some autosampler systems. The primary consideration for the purge is that it be sufficient to maintain essentially 100% oxygen over the surface of the specimen during the test. The use of an excessive purge may increase temperature gradients and heat flow noise.

Specimen Size, Encapsulation and the Use of Covers

There are several effects that relate to the encapsulation and sample size. The first is that, for good OIT reproducibility, the specimen should be distributed in the pan in a reproducible way. Normally the specimen, which consists of a blend of aliphatic components, antioxidants and other additives in a film geometry, is placed in an open, flat bottom pan. When the sample is being heated to the test temperature the specimen melts and flows. At the isothermal test temperature the specimen temperature becomes constant and the steady baseline indicates that chemical and physical processes are constant. One of these processes is the uptake of antioxidant by the oxygen in the specimen environment. The exothermic reaction is initiated after the antioxidant has essentially all diffused to the surface and reacted with the oxygen atmosphere. If the specimen is distributed such that its thickness is uniform over the bottom of the pan, then much of the sample surface will run out of antioxidant at approximately the same time, and the resultant thermal curve will be simple and reproducible. If, because of surface tension, the melted specimen has flowed into a geometry with a thin film on the bottom and a thick meniscus around the edge of the bottom it might be expected that the OIT exotherm will be in two steps reflecting the two extremes of sample thickness. Other factors, that affect the induction process are sample pan shape, surface cleanness, and oxidation (which can effect specimen wetting). Purveyors of this viewpoint have suggested the use of a screen or partial crimp to constrain the sample geometry; however, this approach was not explored in this study.

Another viewpoint is that, at least for some types of samples, at the test temperature the rate of diffusion within the sample is such that antioxidant diffuses from the thick regions into the thin film regions faster than diffusion and reaction at the interface. Thus, the sample geometry may not be important. A mathematical treatment of

Temperature Calibration

For the ASTM E1858 test and the data reported here, the calibration was performed using indium and tin and the manufacturers' two-point calibration software, which also coincides with the ASTM Practice for Temperature Calibration of Diffe4rential Scanning Calorimeters and Differential Thermal Analyzers (E-967) but using a heating rate of 1°C/min. This calibrates the isothermal temperature at 175 and 195°C within 0.1°C. The calibration was performed under the environmental conditions to be used by the test. Thus, the cooling system, purge and pressure were set before temperature calibration. Because of the sensitivity of this test to temperature, this is an important prerequisite to interlaboratory agreement. On a temperature-calibrated power compensation DSC, the program temperature (Tp) and sensor temperature (Ts) are closely related because of the temperature feedback loop. Hence, either the Tp or Ts temperature scale can be used for calibration and readout. (Earlier instrumentation provided only the program temperature; recent instrumentation provides the sensor temperature or both.)

Results

The results of the OIT of motor oil at 3447 kPa using a power compensation pressure DSC are included with the data submitted to the Interlaboratory Test. The results fell well within the range of results from heat flux pressure DSCs; and this data is not further discussed in this paper.

The results of the ambient pressure study of the OIT of HDPE can be seen in Table 1. They include data that was run using exactly the conditions prescribed for the ASTM E37.01.10 Interlaboratory Study (those using 5 mm open capsules) and other data using other commonly used DSC capsules. Each test was run in duplicate with both results reported. The final column shows the average deviation from the duplicate pair. The final table entry gives the results, including average and standard deviation of the original interlaboratory test.

TABLE 1 Results of Ambient Pressure OIT of HDPE

Instrument & type	Encapsulation	Temp(°C)	OIT1(min.)	OIT2(min.)	Ave.(min.)	Ave. Dev.
DSC 7 (P-C)	5 mm open	195.0	24.6	25.9	25.3	0.8
DSC 7 (P-C)	8 mm vented	195.0	29.1	27.2	28.1	0.9
DSC 7 (P-C)	7 mm open	195.0	23.7	28.3	26.0	2.3
DSC 6 (H-F)	7 mm open	195.0	26.3	26.9	26.6	0.3
DSC 6 (H-F)	8 mm vented	195.0	24.9	26.0	25.4	0.5
DSC 6 (H-F)	7 mm open	196.4	20.2	18.6	19.4	0.8
DSC 6 (H-F)	8 mm vented	196.4	17.0	16.4	16.7	0.3
TA2920 (H-F)	5 mm open	195.0	23.5	22.9	23.2	0.3
E37.01.10 IL test	various	195.0			28.3	$\sigma = 4.3$

the diffusion within the sample and across the sample-atmosphere interface should help resolve this question.

In this test a 3 mg high density polyethylene (HDPE) specimen, cut from a film, could be seen (after the test) to have maintained its original geometry in the 7 mm pans; it did not form a meniscus or bead. Thus, the sample thickness appeared to be uniform.

The capsules were all from high-grade aluminum without a deliberate oxide coating. They were either cleaned using the ASTM E1858 test procedure or cleaned by the manufacturer and heated briefly to 375°C in nitrogen to drive off any oil.

A final, but critical, point for use of power compensation DSC is that while open pans are often used for this test, the platinum sample holder lids should *always* be used [2]. These lids form part of the controlled temperature envelope that surrounds the specimen. In the past tThere has been concern when using OIT nitrogen-to-oxygen switchover methods that the oxygen would not diffuse to the surface of the sample. However, because of the low cell volume of power compensation DSCs, the switchover at the specimen surface when the two-hole platinum lids are used is very rapid and at least comparable to that for the larger volume heat flux DSC cells.

Temperature

The temperature for an OIT test is selected to give an induction time between 20 min and 1 h. For the ASTM E37 Interlaboratory Test, these were selected to be 195°C for ambient pressure and 275°C for 3447 kPa data. Data taken inadvertently at 1.7°C higher temperature resulted in OIT values roughly 30% lower. This considerable temperature dependence underlines the need for accurate temperature calibration and minimizing the amount of temperature overshoot.

The ramp rate used for heating to the isothermal test temperature was 40°C/min., which is a small fraction of the fastest rate allowed by the instrument. The reason for selecting this rate was to be consistent with earlier OIT tests which commonly call for a rapid, but controlled ramp. OIT time was set at zero on the arrival of the program temperature to the isothermal step. As shown in Fig. 2, this coincides closely with the arrival of the specimen temperature. Alternatively the time can be reckoned from the start of the heating ramp

In all DSC instruments, the sensor that detects the specimen temperature is outside the pan containing the specimen. It is mounted under, or in, the floor of the sample holder or DSC disk. When a reaction is taking place, when the temperature is being ramped, or during equilibration periods, there are, by necessity, temperature differences between the specimen and the sensor. The temperature accuracy of the DSC depends on having calibrated the system under conditions that are sufficiently equivalent to those being used for the measurement. One factor affecting the temperature calibration is the heating rate. For an isothermal experiment the heating rate is zero. Since calibration is performed by heating through the melting point of a pure material, the calibration should be performed at an appropriately slow scan rate to reduce temperature gradient errors to an acceptable level.

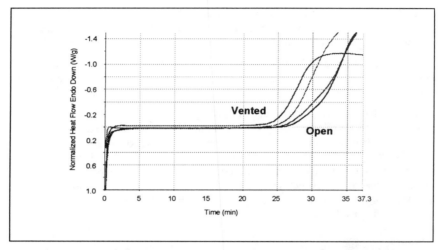

Fig. 3 - OIT of HDPE by Heat Flux DSC6 Using Open and Vented Pans

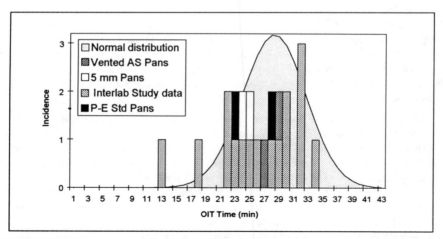

Fig. 4 - OIT Interlaboratory Study Results Plus Additional Power Compensation DSC Data

Discussion

All OIT data obtained for HDPE at the designated temperature of 195°C were within 20% of 28.3 min, the average obtained from the ASTM E37.01.10 Interlaboratory Study. This is relatively good agreement based on results from earlier interlaboratory studies. Almost all the data fell on the low side of this average. The highest values, and those closest to the ASTM E37 average for the power compensation DSC, were those using a 50 ml sealed capsule having vent holes on the side. The good results for this capsule, which is autosampler compatible, indicates that it is not necessary to use a completely open capsule for this analysis. This is also confirmed by the fact that many laboratories are already using this capsule and an autosampling system to completely automate this test. Of course, it is likely that use of this pan, like every other divergence from a standard method, may have a statistical effect on the results. However, it is encouraging to see that the effect, if any, is small. Figure 3 shows the spread of data run in the Perkin-Elmer heat flux DSC6, including data in open and vented pans. Figure 4 graphically presents the data in the table, the data in the original interlaboratory test and the normal distribution curve based on the orignal interlaboratory test data.

Conclusion

In conclusion, while the data taken for this paper has been limited in scope, it is one part of a substantial interlaboratory test whose results are discussed in other papers in this symposium. These data, taken as a whole, show that the OIT test can be performed over a wide range of instruments to obtain comparable data. This particular contribution has focused on the use of a power compensation DSC.

It is hoped that with the additional discussion in this paper that users of this technique will better understand the effect of experimental conditions and the type of considerations to obtain agreement with heat flux instruments. The results from the power compensation DSC have been shown to be comparable to those obtained in a interlaboratory test and also comparable to those obtained with an entry level heat flux DSC. The use of alternative capsules has also been demonstrated. It is hoped that this work will continue to be expanded to extend this demonstration to other samples and test conditions.

References

[1]   Bair, H. E., Polymer Engineering and Science (1973), 13(6), 435-9

[2]   Turi, E. A., Thermal Characterization of Polymeric Materials, Academic Press, New York, 1981, 845-909

[3]   See ASTM E1858 for many references of existing OIT test methods.

[4]   O'Neill, M. J., Analytical Chemistry, (1964), 36(7), 1238-45

Sanford M. Marcus[1] and Roger L. Blaine[1,2]

ESTIMATION OF BIAS IN THE OXIDATIVE INDUCTION TIME
MEASUREMENT BY PRESSURE DSC

REFERENCE: Marcus, S. M., and Blaine, R. L., "Estimation of Bias in the Oxidative
Induction Time Measurement by Pressure DSC," Oxidative Behavior of Materials by
Thermal Analytical Techniques, ASTM STP 1326, A. T. Riga and G. H. Patterson, Eds.,
American Society for Testing and Materials, 1997.

ABSTRACT: Oxidative induction time (OIT) is defined as the time to the onset of
oxidation of a test specimen exposed to oxygen at an elevated isothermal test temperature.
In the pressure differential scanning calorimetry embodiment of this test, the test specimen
is often exposed to the oxidizing atmosphere as the apparatus is heated from ambient to
the isothermal test temperature. This creates a bias in the measurement due to undetected
oxidation on heating. An expression, based upon the Arrhenius equation, is derived and
then numerically integrated to obtain an estimate of the bias introduced into the OIT
measurement. The bias is dependent on the activation energy of the reaction and on the
heating rate. It is found to be less than 1.2 minute for the most common heating rates, and
is less than 3 minutes for the most extreme sets of experimental conditions. The bias is
small when compared to experimental repeatability of the OIT measurement and to the
mean of OIT values. For this reason, it may be ignored in all but the most extreme cases
of low activation energy, very slow heating rates, very low OIT values and high test
temperatures.

KEYWORDS: bias, differential scanning calorimetry, edible oils, lubricants, oxidative
induction time, oxidative stability, polyolefins, pressure differential scanning calorimetry,
thermal analysis

---

[1] Applications chemist and applications development manager, respectively, TA
Instruments, Inc., 109 Lukens Drive, New Castle, DE 19720.

[2] Corresponding author.

Introduction

Oxidative induction time (OIT) is defined as the time to the onset of oxidation of a test specimen exposed to an oxidizing gas at an elevated test temperature. OIT values are used as an indexes to estimate the relative stability of materials to oxidation. They are typically used as quality control tools and to rank the effectiveness of various oxidation inhibitors added to hydrocarbon products such as polymers [1, 2] , lubricating oils and greases [3] as well as edible fats and oils.

OIT measurements are commonly carried out using differential scanning calorimetry by temperature programming the specimen in an open sample pan from ambient to an isothermal test temperature under an inert atmosphere.   Once the test temperature is reached and equilibrium is established, the purge gas is changed to an oxidizing gas (i.e., air or oxygen) and the clock is started. The elapsed time from the first oxygen exposure (t_o) to the onset of the oxidation (t_{onset}) is taken as the OIT value (see Figure 1). Oxidation is observed as an exothermic peak, although in most OIT experiments, only enough of the peak is recorded to determine its onset.

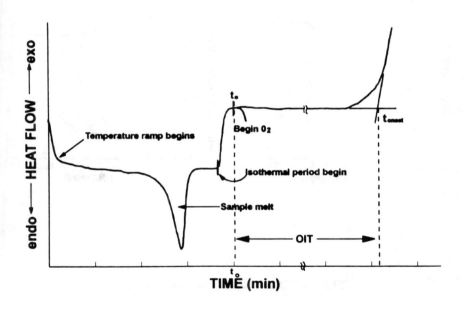

FIG 1-- OIT experimental curve.

The temperature program used to achieve the OIT test conditions is made up of two parts as seen in Figure 2A. The first is the constant rate of temperature rise from ambient to the test temperature. The second portion is the establishment of the isothermal test temperature (T_t) which is maintained until the experiment is terminated. The ramp rate to the isothermal test temperature is set as high as possible (to reduce analysis time) without overshooting the test temperature. The isothermal test temperature is typically selected to produce OIT values between 15 and 100 minutes; the higher value to limit analytical time and increase productivity, the lower value to limit the effect of measurement imprecision. Typical isothermal test temperatures range between 150 and 210°C [1, 2, 3, 4]. The higher the test temperature, the lower the OIT value. The default for most OIT measurements is 200°C, with minor temperature adjustments being made depending upon the material being tested.

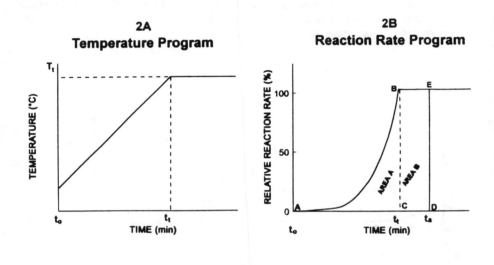

FIG 2 -- OIT temperature reaction rate profile.

If a material is strongly stabilized, its OIT value may be quite long. In an attempt to reduce analytical time, the test temperature is sometimes elevated. Unfortunately, many oxidation inhibitors have an appreciable volatility at temperatures above 180°C and so increasing the test temperature results in the volatilization of the additives rather than in a measure of their chemical effectiveness. However, the OIT measurement may be accelerated at lower test temperature through the use of elevated partial pressures of oxygen. Going from the 21% oxygen content of air as a reactant to 100% oxygen is one alternative. Another is through oxygen atmospheres at elevated pressures making use of pressure DSC (PDSC).

Using PDSC, the OIT test is conducted under slightly different experimental conditions than those used with the standard DSC. In PDSC, the specimen is pressurized with the reactive oxygen gas at room temperature, followed by temperature programming to the test temperature at a constant rate. The initial time for the OIT measurement (t_i) is taken when the programmed temperature reaches the test temperature (T_t).

These modified experimental conditions (exposing the sample to oxygen at the start of the temperature ramp conditions) raises the question, "How much oxidation takes place as the sample is programmed from ambient to the test temperature?" This early oxidation results in an underestimation of the OIT value at the isothermal test temperature and represents a measurement bias.

Theory

In kinetic expressions, such as those governing the OIT measurements, the relationship between the kinetic rate constant (k) and the temperature (T) is given by the Arrhenius expression.

$$k(T) = Z \exp(-E/RT) \tag{1}$$

where:

$k(T)$ = specific rate constant at temperature T (1/min),

Z   = pre-exponential factor (1/min),
E   = activation energy (J/mol),
R   = molar gas constant (= 8.3143 J/mol K), and
T   = absolute temperature (K).

The effect of the Arrhenius expression on the OIT measurement may be seen in Figure 2B, which displays the relative reaction rate (the ratio of the reaction rate at a given temperature to that at the test temperature [i.e., $k(T) / k(T_t)$]) on the ordinate versus time on the abscissa for the same region displayed in Figure 2A. The reaction rate is small at low temperatures and increases exponentially with temperature until the test temperature is reached. Once the test temperature is achieved, the reaction rate is assumed to be constant. (This model neglects, however, that some samples (e.g., polyolefins) undergo melting between ambient and the test temperature and that the reaction rate is different in the crystalline than in the amorphous form [5] .)

The amount of reaction which has taken place between ambient and the test temperature is a bias underestimating the OIT value. This bias is represented in Figure 2B by the Area A bounded by points ABCA. This area is obtained by integrating equation (1) over the limits from the time of first exposure to oxygen (t_o) to the time the test temperature is reached (t_t); that is, the interval of the temperature ramp.

$$\text{Area A} = \int k(T) \, dt = Z \int \exp(-E/RT) \, dt \qquad (2)$$

For the constant heating rate region, the heating rate (β) provides the relationship between changing temperature (dT) and changing time (dt); $dt = dT / \beta$. Substituting this relationship into equation (2) yields:

$$\text{Area A} = (Z/\beta) \int \exp(-E/RT) \, dT \qquad (3)$$

A second area, Area B, is described in Figure 2B, which corresponds to the closed area BCDE. This area represents the amount of reaction which takes place at the test temperature over the time interval from t_t to t_a. This interval is selected so that Area B is equal to Area A. When this is done, the time t_a - t_t range is then equal to the bias in the OIT measurement due to the pre-test temperature reaction and is given the symbol ΔOIT. Area B is equal to the reaction rate at the test temperature (T_t) multiplied by ΔOIT.

$$\text{Area B} = k(T_t) \ \Delta\text{OIT} = Z \exp(-E/RT_t) \ \Delta\text{OIT} \qquad (4)$$

Setting equations (3) and (4) equal to each other and solving for ΔOIT:

$$\Delta\text{OIT} = [\int \exp(-E/RT) \, dT \ ] / [\beta \exp(-E/RT_t)] \qquad (5)$$

Values of the isothermal test temperature (T_t) and activation energy (E) parameters are needed to solve equation (5). Many OIT tests are carried out at 200 °C (= 473 K). This value is selected for the value of T_t in this evaluation. The second parameter, activation energy, is determined from a series of OIT measurements of the same material at several isothermal test temperatures. According to equation (6) (derived by taking the logarithmic form of equation (1) and substituting the value of ln k = -ln OIT + constant from the general rate equation at constant conversion), a plot of natural logarithm of the OIT versus the reciprocal of the absolute isothermal test temperature produces a straight line, the slope of which is equal to E/R.

$$\ln OIT = [(E/R) \times 1/T] + \text{constant} \qquad (6)$$

Figure 3 illustrates such a data treatment for a polyethylene sample tested under 3.5 MPa (500 psig) pressure of oxygen, using a PDSC cell.

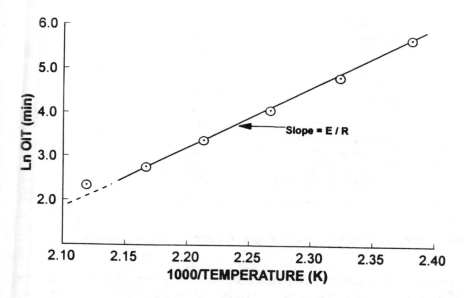

FIG 3 -- Determination of activation energy.

Activation energy values for OIT measurements for three common classes of materials (edible oils, polyolefins and lubricants) to which the OIT procedure is applied are collected from the literature in Table 1. The values for E for these materials range from a low of 70 (for edible oils) to a high 250 kJ/mol (for polypropylene) with a value of 130 kJ/mol as an average.

With appropriate values for T_t and E selected, equation (5) may be evaluated. For convenience, the exponential terms, $\int \exp(-E/RT)\, dT / \exp(-E/RT_t)$, are collected into a term, called the Reaction Fraction (F), which has the units of temperature. Equation (5) then takes the form:

$$\Delta OIT = F / \beta \qquad (7)$$

TABLE 1 -- Typical OIT activation energies.

Material	Activation Energy (kJ/mol)
Corn Oil [6]	99
Rapeseed Oil [6]	72
Soybean Oil [6]	67
Sunflower Oil [6]	85
Peanut Oil [7]	110
Polyethylene [8]	89
Polyethylene [9]	165
Polyethylene [10]	192
Polyethylene, Crosslinked [11]	144
Polypropylene [10]	249
Lubricating Oil[3] [12]	121
Lubricating Grease[3] [13]	146

[3] Pressure DSC at 3.4 MPa (500 psig) oxygen. All others at ambient pressure.

The evaluation of F requires the integration of the exponential term in the numerator. This is difficult to do exactly but may be estimated using numerical integration techniques. The numerical integration process used, known as Simpson's Rule, is illustrated in Figure 4 where the relative reaction rate is plotted as the curved line with temperature as the abscissa. The area under the curve may be estimated using a series of rectangles with a temperature width 2°C multiplied by the reaction rate constant for the midpoint temperature for that 2°C range. For example, the area BCFG may be estimated by the rectangle HCFI. The rectangle's over estimation of the area JGI below the midpoint temperature approximates the area underestimation BHJ above this value. By summing all of the individual rectangles from ambient to the test temperature, an estimation of the value for the fraction F may be obtained.

Table 2 shows a series of Reaction Fraction values calculated for representative activation energies covering the range of interest for OIT studies. With these values in hand, the effect of heating rate (β) on the value of the OIT bias may be estimated from equation 7.

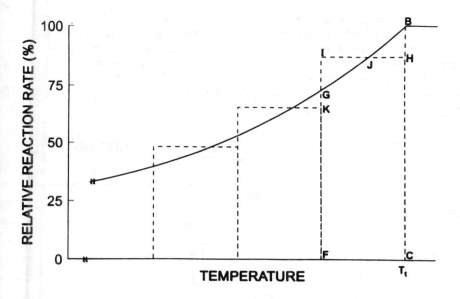

FIG. 4 -- Numerical integration.

TABLE 2 -- Effect of activation energy on reaction fraction
(T_t = 200 °C = 473K)

Activation Energy (kJ/mol)	Reaction Fraction (F) (K)
70	24.01
100	17.29
130	13.51
150	11.79
200	8.94
250	7.20

Discussion

Table 3 presents ΔOIT values for common heating rates for 70, 130 and 250 kJ/mol, the minimum, average and maximum activation energy values of Table 2. These data show that the OIT bias is smallest for high activation energy materials (such as polypropylene) and high heating rates. It is largest for the lower activation energy materials (such as the edible oils) and low heating rates. For the typical heating rate of 20°C, the bias due to oxygen exposure at the start of the OIT experiment, is less than 1.2 minutes. The importance of this 1.2 minute maximum bias may be assessed by comparing it first to the mean value of OIT measurements and secondly, to the precision of the OIT measurement itself.

TABLE 3 -- Effect of heating rate and activation energy on OIT bias
(test temperature = 200°C)

Activation Energy (kJ/mol)	Heating Rate (°C/min)	OIT Bias (min)
70	5	4.8
	10	2.4
	20	1.2
	40	0.6
130	5	2.7
	10	1.4
	20	0.7
	40	0.3
250	5	1.4
	10	0.7
	20	0.4
	40	0.2

The Pressure DSC measurement of OIT is ordinarily applied to very stable materials. In the geosynthetic industry where the OIT measure is made on waste pit liners, the value for OIT is typically 300 to 1000 minutes [4]. Clearly, a bias of 1.2 minutes in such a measurement is trivial. More commonly, however, the OIT value for other materials is between 15 and 100 minutes. For the shorter of these time periods, the bias begins to be significant. In cases of low OIT values, a second comparison is made to repeatability of the OIT measurement.

ASTM standard test methods, including those for OIT, contain repeatability (within laboratory precision) information derived from interlaboratory test programs. These results show that repeatability tends to be relatively constant for a particular class of materials and is only a weak function of the OIT value itself. The best obtainable precision for OIT values are ca. 2 minutes while the pooled standard deviation for a variety of materials is around 5 minutes [1, 2, 3, 13]. Comparing the repeatability of the OIT test with the bias induced by exposing the test specimen to oxygen from ambient temperature, shows the bias to be smaller than obtainable precision.

Conclusions

The bias introduced into the OIT measurement by exposing the test specimen to oxygen at room temperature and then heating to the test temperature is found to be small when compared to the experimental repeatability and to the mean values of the OIT values. For this reason, it may be ignored in all but the most extreme cases of low activation energy, very slow heating rates, and high test temperatures. The ease-of-use benefits, achieved by exposing the test specimen to the reactive gas from the start of the experiment are likely to out weight the small bias effect in most OIT measurement whether under Pressure DSC or standard DSC condition.

Extension to Isothermal Crystallization and Isothermal Kinetics

This same procedure may be used with other types of experiments where the temperature is linearly raised to a specific test temperature where the time to a thermal event is measured. Two additional examples include polymer isothermal crystallization and isothermal kinetics such as thermoset cure. In isothermal crystallization, the test temperature is approached from higher temperatures. For isothermal kinetics, the test temperature is approached from lower temperatures as with the OIT measurement.

TABLE 4 - Activation energy for isothermal crystallization [13]

Materials	Activation Energy (kJ/mol)
Polyethylene	88
Nylon 6	357
PEEK	472
PET	1080

Table 4 lists some of the activation energies for isothermal crystallization reactions determined in our laboratory using the Sestak Berggren equation [14, 15], which is similar in form to the general rate equation . Polyethylene is considered to be a "fast" crystallizing material with its activation energy of 88 kJ/mol while poly(ethylene terephthalate) is considered to be a slow crystallizer at 1080 kJ/mole. Comparing these activation energies to the bias values listed in Table 3, shows that, except for the very lowest of activation energies, even for modest temperature rates of cooling (e.g., 10 °C/min) this procedure may be used for isothermal crystallization with a bias of less than 1 minute.

For thermoset cure reactions, activation energies are often in the range of 60 to 100 kJ/mol. These values are low compared to most OIT and polymer crystallization reactions. By comparison to Table 3, such low activation energy curing reactions benefit from a very rapid temperature rise to the test temperature.

REFERENCES:

[1]   ASTM D3895-94, "Test Method for Oxidative Induction Time of Polyolefins by DSC", American Society for Testing and Materials.

[2]   ASTM D4565-94, "Test Method for Physical and Environmental Performance Properties of Insulations and Jackets for Telecommunications Wire and Cable", American Society for Testing and Materials.

[3]   ASTM D5483-93, "Test Method for Oxidation Induction Time of Lubricating Greases by Pressure DSC", American Society for Testing and Materials.

[4]   D5885, "Test Method for Oxidative Induction Time of Polyolefin Geosynthetics by High Pressure Differential Scanning Calorimetry", American Society for Testing and Materials.

[5]   Yelin, C.F.; Anal. Calor., Vol. 4 (R.S. Porter and J.F. Johnson (Eds.), Plenum Press, 51-66 (1977).

[6]   Kowalski, B.; Thermochim. Acta, 156, 347-358 (1989).

[7]   Flynn, J.H.; Meeting of Penn. Manuf. Confectioners Assoc. (1985).

[8]   Thomas, R.W,; Ancelet, C.R.; Proc. Geosynthetic Conf., 2, 915-924 (1993).

[9]   Howard, J.B.; Gilroy. H.M.; Polym. Eng. Sci., 15 (4), 268-271 (1975).

[10]   Marshall, D.I.; George, E.J.; Turnipseed, J.N.; Glenn, J.L.; Polym. Eng. Sci., 13 (6), 415-421 (1973).

[11]   Kramer, E.; Koppelmann, J.; Polym. Degrad. Stab., 16, 261-275 (1986).

[12]   Walker, J.A.; Tsang, W.; Soc. Autom. Eng. Tech. Paper 801383, (1980).

[13]   Foreman, J.A.; Blaine, R.L.; 41st Ann. Tech. Conf., Soc. Plast. Eng., 2, 2409-2412 (1995).

[14]   Sestak, J.; Berggren, G.; Thermochim. Acta, 3, 1 (1971).

[15]   Gorbatchev, V.M.; J. Therm. Anal., 18, 194 (1980).

Rolf Truttmann[1], Kristy Schiano[1], and Rudolf Riesen[2]

OXIDATIVE INDUCTION TIME (OIT) DETERMINATIONS OF POLYETHYLENES: INFLUENCES OF TEMPERATURE, PRESSURE, AND CRUCIBLE MATERIALS ON THE RESULT

REFERENCE: Truttmann, R., Schiano, K., and Riesen, R., "**Oxidative Induction Time (OIT) Determinations of Polyethylenes: Influences of Temperature, Pressure, and Crucible Materials on the Result,**" Oxidative Behavior of Materials by Thermal Analytical Techniques, ASTM STP 1326, A. T. Riga and G. H. Patterson, Eds., American Society for Testing and Materials, 1997.

Abstract: The standard procedures for the oxidation induction time (OIT) determinations have been developed to get reproducible and comparable results on the same sample by different users. Experimental conditions such as isothermal measuring temperature, pressure of the surrounding oxygen concentration, and type of crucible material are prescribed to avoid large variations. Many difficulties met in the past may be due to improper selection of the mentioned parameters, namely the temperature accuracy and crucible type and cleanliness. Because of the autocatalytic behavior of the oxidation, the accurate temperature calibration is of tremendous importance, since usual calibration is done dynamically, but the measurements are performed isothermally, that is, without dynamic lag between sample and furnace. Therefore, the shift of the OIT by the variation of temperature, pressure (oxygen concentration), and crucible material is investigated.

Keywords: oxidation induction time, ASTM, polyethylene, polyolefins, influences of temperature, pressure, crucible material

---

[1] Product Management, Mettler-Toledo, Inc., Hightstown, NJ 08520-0071.

[2] Manager, Market Support Materials Characterization, Mettler-Toledo AG, Analytical, CH-8603 Schwerzenbach, Switzerland.

INTRODUCTION

Oxidative induction time (OIT) is a relative measure of the degree of oxidative stability of the material evaluated at the isothermal temperature of the test. For many years, OIT standards exist in various forms, for example, the ASTM Test Method for Oxidative Induction Time of Polyolefins by Differential Scanning Calorimetry (D 3895), the ASTM Specification for Polyethylene Plastics Pipe and Fittings Materials (D 3350) and the new European standard on Plastics piping and ducting systems - Polyolefin pipes and fittings - Determination of oxidation induction time (prEN728:1993). The presence or effectiveness of antioxidants may be determined by these or related methods. The OIT values thus obtained may be compared from one hydrocarbon to another, or to a reference material to obtain relative oxidative stability information. Typical materials investigated are polyolefins, lubricants and greases, edible oils and fats.

The rate of oxidation and moreover the OIT is influenced by many parameters, for example, the temperature and partial oxygen pressure of exposure as well as the selection of crucible material, gas diffusion, exposed surface area, and even by the mode of result evaluation.

The scope of this article is to give an insight of the many influences. The understanding of these relations is important for optimal design and reproducibility of the measurements. A large variety of materials stabilized for many different applications, also show a large band width of OIT (Fig. 1).

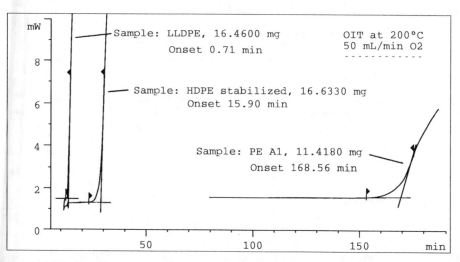

Fig. 1: OIT of various polyethylene samples measured at 200 °C using the procedure prEN 728, but allowing for slightly increased sample masses. The onset values are the time from the start of oxygen exposure. The diagram axis displays the total experimental time.

But the same sample may show a large range of OIT due to nonhomogeneous material, when specimens are taken from different pellets (Fig. 2).

It is often difficult to decide whether the variations of OIT values found during investigation of new samples are due to the variation of the material itself or just a result of other influences. This makes rigorous control of the experimental parameters most important and knowledge of the above mentioned influences is a prerequisite for the correct interpretation of the results and the necessary conclusions about the oxidative stability.

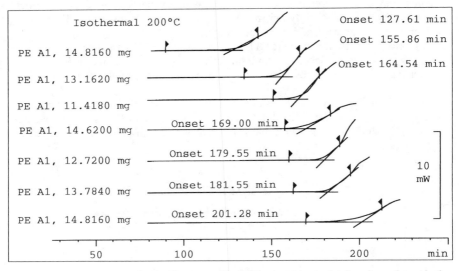

Fig. 2: OIT determinations of various species of the same sample showing a large in-homogeneity throughout the material. Method used: prEN 728 at 200 °C. The time axis displays time of oxygen exposure.

INFLUENCES ON OIT

As already mentioned above, OIT values are strongly influenced by many parameters. They may be categorized as follows:

A: Influences by the sample material itself
B: Influences by experimental parameters

Category A is usually the objective of investigation for a given or optimal set of B. Category A summarizes the dependence of the nature of the main material (for example, polymer, olefin) and additives, fillers, inhibitors, catalysts.

Category B comprises the controlling parameters as follows:
- temperature and the instrument calibration;
- oxygen pressure or concentration: air, pure oxygen, pressurized cells;

- diffusion of oxygen to the specimen and oxidized products to the atmosphere: design of the DSC cell, purge gas flow rate, and specimen holder or crucible;
- sample amount and shape; and
-     evaluation procedures.

This is the reason why all these parameters are usually fixed in the standards. The time necessary to increase the oxygen to an effective concentration after switching over from nitrogen purge may also be important.

EXPERIMENTAL PROCEDURES

Samples: Polyethylene in pellet or strip shape of various origin. To reduce the influence by the sample, the same black stabilized PE (Hostalen GM5040T12), which was confirmed by the manufacturer to be homogeneous from pellet to pellet, has been used for most of the comparison measurements. These pellets have been cut horizontally into two flat pieces to give approximately cylindrical shapes of 4 mm diameter and 1 mm thickness. During the analysis, the PE melted and covered the bottom of the pans (diameter 6 mm) totally.

ASTM round robin reference materials, oil (C) and polyethylene (D) have been used for checking the influence of oxygen pressure [1].

Instrumentation: Mettler Toledo DSC821^e with the STAR^e software, if not otherwise specified. For routine measurements, the system is fitted with sample changer and gas controller to activate the right gas flow for each segment. Special crucibles have been inserted manually at 50 °C. The reference crucible is always of the same type as the sample crucible but empty.

OIT procedures: The mentioned standard [1] was used, including calibration of the measuring cell using indium and tin at a rate of 1 K/min. But, the sample masses deviate significantly from the given standard; therefore, all masses of the specimen are mentioned. The masses used are closer to the sample masses specified in prEN 728. One of the important differences between the standards prEN 728 and in Ref. [1] can be found in the sample weights, that is, 10 to 14 mg compared to 3.0 to 3.3 mg, respectively. The automated STAR^e system allows for termination of the experiment. For example, termination occurs when the exothermic DSC signal reaches 3 mW followed by a waiting time of 10 min to pass the exothermic maximum. Therefore, the curves end at different times.

Determination of the start of the oxidation (OIT): The OIT value is the time from the switch-over nitrogen to oxygen to the extrapolated onset. This onset is defined by the intersection of the extrapolated baseline with the tangent to the inflection point. Another sometimes used procedure is to specify a threshold value above the baseline, where the tangent to the DSC curve is drawn (instead of the inflection tangent).

All diagrams with the DSC curves display the exothermic direction upwards.

RESULTS

Most of the subsequent comparisons have been performed on a relative base, that is, keeping the experimental parameters as constant as possible. As a reference for the following comparisons, the measurements at 210 °C in 40 μL aluminum crucibles, using the standard [1] with weights of 10 to 13 mg, have been selected. These measurements show a mean OIT value of 35.2 min with standard deviation 1.4 min over five measurements (see also Fig. 7).

The influence of gas flow rates up to 80 mL/min on the onset of fusion of a pure metal is less than 0.1 °C, that is, in the same range as the reproducibility of the melting point determination.

Temperature Influences

The temperature influence on the oxidation reaction and its auto-acceleration behavior is quite critical. One or two degrees variation affect the OIT much more than variations seen by the reproducibility of the reference conditions (Fig. 3). Hence, for comparison purposes between labs, the temperature calibration procedure (including the reference material) must be the same.

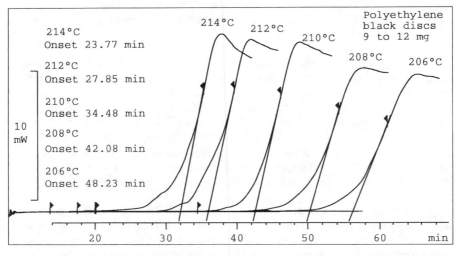

Fig. 3: Influence of temperature on OIT values by a series of 2 °C intervals between 206 and 214 °C, using the ASTM method but with masses between 9 and 12 mg of the black polyethylene pellets. The onset values printed are OIT values (oxygen exposure); the abscissa displays the total experimental time (the nitrogen purge lasts 8 min).

Pressure Influences

In cases in which the sample is volatile or contains volatiles, for example, lubricants or oils, pressurized DSC cells may be used to suppress vaporization before the OIT value is determined. The increase of the pressure also increases the oxygen concentration above the sample and, therefore, increases the rate of oxidation. This allows the application of lower temperatures for the measurement while still giving reasonable OIT values between 20 and 60 min (Fig. 4).

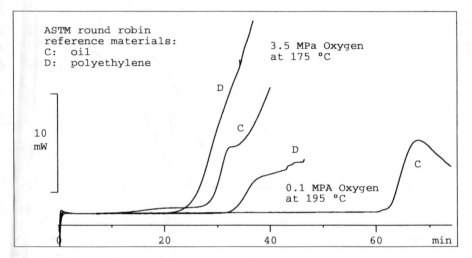

Fig. 4: Influence of oxygen pressure on OIT values: For both types of materials (oil, C, and polymer, D) the OIT value is drastically reduced by increasing the pressure by a factor of 35, even though the temperature was lowered by 20 °C. The time axis displays time of oxygen exposure only. Measurements performed using METTLER DSC27HP.

Hence, for the same material, pressure may be increased but temperature must be lowered to get OIT values within an hour measurement duration. At lower oxygen concentrations, the resistance against oxidation better resolved than at high concentrations, that is, the OIT values are further apart to each other. If the oxygen concentration is reduced more by dilution with nitrogen(that is, using air), the OIT values may overpass the given limits of 2 h, for example.

Influence of Crucibles

The nature of the crucible material may have a catalytic effect. For example, this is sometimes intended to simulate the contact of polymer and copper as in the insulation of copper wires. Figure 5 compares the various crucible materials under the same conditions. The crucibles made of copper or aluminum have exactly the same shape (6 mm diameter, 1.5 mm height, volume of 40 μL), but the pans made from alumina or platinum are a little taller with 4.3 mm and a diameter of 5.6 mm (volume of 70 μL). Therefore, for very careful studies, very well-cleaned glass crucibles may be applied.

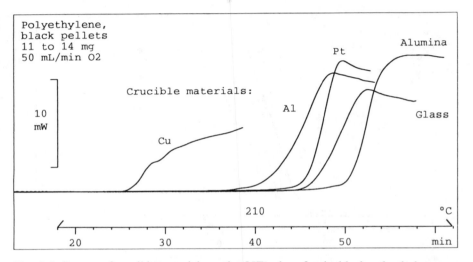

Fig. 5: Influence of crucible materials on the OIT values for the black polyethylene, measured following the procedure [1], but with more material to cover fully the bottom of the crucibles.

Influence of Copper Contacts

Small amounts of copper (or other catalytic metals) will shorten the OIT (Fig. 6). A critical point in these studies is the amount of copper in contact with the species. The OIT without copper is shortened to about half of its value when using a copper pan instead of an aluminum pan (Fig. 6). To demonstrate intermediate steps, the PE samples were contacted in different ways with copper (Fig. 6).

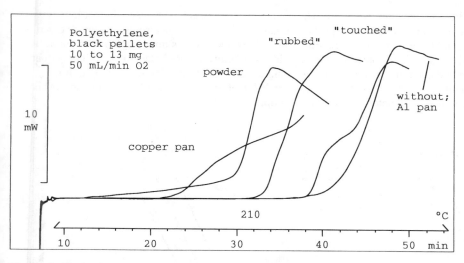

Fig. 6: Oxidation induction time influenced by copper contact: the more copper present the smaller the OIT: All specimens but one have been measured in aluminum pans. The following treatments were applied to the upper surface of the sample:
powder: covering with 0.3 mg of copper powder,
"rubbed": strongly frictioning a copper plate,
"touched": slightly frictioning a copper plate, and
copper pan: no special treatment, specimen placed in copper pan directly.

Influence of Evaluation Procedure

Because of stabilizers or other reasons, the measuring curves are often not perfectly S-shaped as expected. Therefore, it is matter of definition which point of the DSC curve is being used to get the extrapolated onset. Hence, the OIT value can be reduced by more than 10% if a small threshold value is selected to draw the tangent, instead of taking the inflection tangent (Fig. 7). This procedure gives more weight to the starting oxidation, that is, to the initial deflection from the baseline. For comparison purposes, it is again very important to also specify the evaluation procedure or any changes to it.

CONCLUSION

Many investigations resulted in standard procedures to determine the oxidative properties of hydrocarbons by DSC to allow comparison of influences of the nature of the material, additives, and aging processes. The oxidation behavior can be measured in various ways. The most often used method is the oxidation induction time determination under given isothermal conditions. As the investigations show, the result is influenced by the temperature and partial oxygen pressure, the selection of crucible material, the contact of copper catalyst and even by the means of evaluating the results. Despite the thoroughly defined conditions in standard procedures, knowledge of the sensitivity of the results to the above mentioned parameters are essential for optimal analysis.

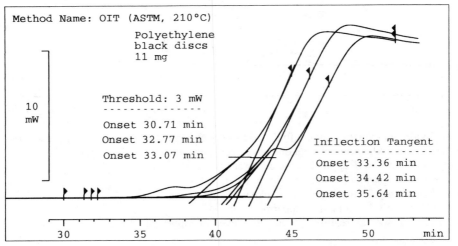

Fig. 7: OIT measurements of black polyethylene at 210 °C. Applying different evaluations (inflection tangent or tangent at a threshold value of 3 mW), different OIT values are gathered.

REFERENCES

[1]   Project TM-01.10A of ASTM Committee E37, Standard test method for determining oxidation induction time of hydrocarbons by differential scanning calorimetry.

Roger L. Blaine[1] and Mary B. Harris[1]

A PROPOSED REFERENCE MATERIAL FOR OXIDATIVE INDUCTION TIME BY DIFFERENTIAL SCANNING CALORIMETRY

REFERENCE: Blaine, R. L. and Harris, M. B., **"A Proposed Reference Material for Oxidative Induction Time by Differential Scanning Calorimetry,"** Oxidative Behavior of Materials by Thermal Analytical Techniques, ASTM STP 1326, A. T. Riga and G. H. Patterson, Eds., American Society for Testing and Materials, 1997.

ABSTRACT: A polyethylene film sample, inhibited with a hindered phenol antioxidant, is proposed as a Standard Reference Material for oxidative induction time (OIT) testing. The mean OIT values, derived from nine interlaboratory studies and for a number of experimental conditions, are presented. The material is found to be statistically homogeneous, a necessary condition for a reference material. Further, the effects of temperature, oxygen pressure, and storage time on the proposed reference material are explored. As a kinetic parameter, the OIT value appears to be decreasing with time but in a well behaved and predictable manner. The use of a table and graph permit the user of the material to estimate its OIT value in the future. Because the material has been thoroughly tested in a wide variety of OIT conditions, it appears to be the best currently available candidate and is offered for consideration as an OIT Reference Material.

KEYWORDS: calibration, differential scanning calorimetry, oxidation, oxidative induction time, oxidative stability, reference materials, thermal analysis

Oxidative induction time (OIT) is a widely used parameter for the oxidative stability of polymers, edible oils, and lubricants. It is typically used as a quality control tool and to rank the effectiveness of various oxidation inhibitors added to hydrocarbon products. OIT is defined as the time to the onset of oxidation of a test specimen exposed to an oxidizing

---

[1] Applications development manager and applications chemist, respectively, TA Instruments, Inc., 109 Lukens Dr., New Castle, DE 19720.

gas at an elevated isothermal test temperature. OIT is a kinetic parameter (that is, one dependent on both time and temperature) and is not a thermodynamic property.

The analytical precision and mean value for the OIT determination are known to depend on a large number of experimental parameters including isothermal test temperature, specimen mass and surface area, purge gas flow rate, and catalytic impurities [1]. Because of these effects, it is quite common for laboratories to get widely different OIT values when testing the same material. Interlaboratory correlation of results are likely to improve with the use of an OIT Reference Material of known characteristics to serve as a performance standard.

According to ISO Guide for Certification of Reference Materials - General and Statistical Principles, a good reference material has a number of desirable properties including a well-documented analytical value, homogeneity, stability, ready availability and traceability to a National Reference Laboratory (NRL). Unfortunately, an OIT Reference Material is not ideal since, by its nature, it does not meet all of these criteria. For example, OIT is not a thermodynamic property and is therefore not easily made traceable to a NRL. Further, OIT is a kinetic property so its value will likely change with time and therefore lacks stability.

Background

Over the last few years, a number of standard test methods for the OIT measurement have been developed, each with its own set of experimental conditions, aimed at optimizing the test for specific products (see Table 1). In each method, intra- or interlaboratory studies or both were conducted to test for ruggedness and provide a precision and bias statement for the standard test method.

One of the most thorough studies was that of ASTM Committee D9 on Electrical Insulation Materials which, in 1994, revised the OIT section to the ASTM Test Method for Physical and Environmental Performance Properties of Insulations and Jackets for Telecommunications Wire and Cable Materials (D 4565). In the development of that revision, five interlaboratory studies were conducted in a series of ruggedness tests to explore one or more of the experimental parameters. As part of that work, a single test sample of polyolefin film was included at each stage of the work as an internal reference material. Over the four years of the test program, a very large amount of data was generated on this single material.

At the end of the ASTM Committee D9 test program, it was proposed that the polyethylene film internal reference material be considered as a Standard Reference Material for the OIT test. The National Institute of Standards and Technology (NIST) and ASTM were contacted to see if they would care to take on the responsibility for distribution of the material as an OIT Standard Reference Material. Both of these

TABLE 1 - ASTM standard methods for oxidative induction time.

Method	Title
D 3350	Specification for Polyethylene Plastics Pipe and Fittings Materials
D 3895	Test Method for Oxidative Induction Time of Polyolefins by Differential Scanning Calorimetry
D 4565	Test Methods for Physical and Environmental Performance Properties of Insulations and Jackets for Telecommunications Wire and Cable
D 5483	Test Method for Oxidation Induction Time of Lubricating Greases by Pressure Differential Scanning Calorimetry
D 5885	Test Method for Oxidative Induction Time of Polyolefin Geosynthetics by High Pressure Differential Scanning Calorimetry
E 1858	Test Method for Oxidative Induction Time of Hydrocarbons by Differential Scanning Calorimetry

organizations declined this offer because of the need for additional testing, the relatively small market opportunity, and instability of the reference material.

The importance of this material was recognized by TA Instruments, and a large quantity of the polyolefin film was purchased from the original manufacturer. Since that original purchase, the material has been protected for future use, and additional testing has been performed. The material has been tested for key properties of homogeneity and stability and has been used in several additional interlaboratory studies.

It is the purpose of this paper, then, to collect the large amount of experimental information on this material and to propose the material as an OIT Standard Reference Material.

Material

The proposed OIT Reference Material is in the form of translucent film 0.22 mm (9 mils) in thickness. It is a high density polyethylene film, lightly stabilized with Irganox ®1010, a hindered phenol antioxidant package. Its melting profile, as determined by ASTM

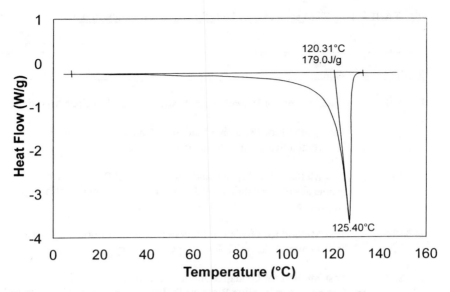

FIG. 1 -- Polyethylene reference material melting profile.

Standard E 794 Test Method for Melting and Crystallization Temperatures by Thermal Analysis, is shown in Fig. 1 and has a peak melting temperature of 125°C. After manufacture, the film was warehouse stored in the dark as a roll. Before examination and preparation of the reference materials kits, the outer layer of this roll was withdrawn and discarded

The proposed reference material is packaged as two 8- by 13-cm sheets of the film, enough for more than 500 individual OIT measurements. Since antioxidant packages can be susceptible to migration between adjacent leaves of material, the two sheets are stored in an envelop of the same film which serves as a sacrificial barrier. These materials are then stored in an opaque polyolefin zip-lock bag to shield them further from environmental exposure.

Homogeneity

Before packaging, the roll of film was partitioned into a series of sections across the breadth and length for homogeneity testing of the antioxidant distribution. A single laboratory obtained OIT values using ASTM Test Method D 4565. The mean OIT resulting from this homogeneity testing, with 48 total determinations, was 30.0 min with a standard deviation of ±1.2 min.

Figure 2 shows a histogram distribution of the OIT values determined in this homogeneity study. The number of OIT values within each 0.5-min range is displayed on the ordinate versus the OIT value on the abscissa. The shape of the curve is one indication of the

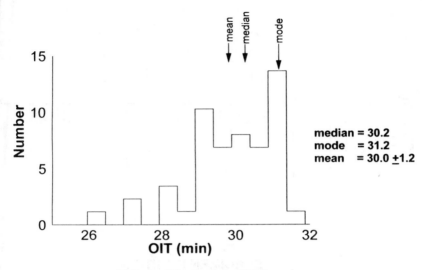

FIG. 2 -- Distribution of OIT results.

randomness of the experimental data with a "bell shaped" distribution resulting if it is truly "normal."

The distribution of results in Fig. 2 appears to be somewhat skewed toward lower values. The most probable mode value for the distribution is 31.2 min with the mean and median at 30.0 and 30.2 min, respectively. The mean is lower than the mode by 1.2 min (4%). Moreover, the mode is very close to the maximum value of 31.6 min.

This distribution of results was statistically tested for skewness using the Pearson's index. The Pearson's skewness index for this distribution is -0.39 indicating a slight skew toward to lower values. However, only indexes greater than ± 1.0 are statistically significant in indicating skew [2]. Therefore, the data may be considered to have a normal distribution and to be amenable to gaussian statistics.

Figure 3 presents another graphical tool to test for homogeneity. Here the same OIT values are presented on the ordinate as a function of their juxtaposed position on the abscissa. The values are plotted around the mean value of 30.0 as an aid to the eye. The data may be visually inspected to detect whether there are any noticeable trends in the data or if there are clusters of measurements of the same OIT values. Such trends or clusters might exist, for example, if the antioxidant package is not well dispersed. Except for a cluster of four measurements near position 30 and another near position 45, the data appears to be quite random with regard to both magnitude and position, above or below the mean.

The data was statistically examined for trends or clusters using the Shewhart rational subgroup technique [3]. In the Shewhart test, the data is partitioned into a series of

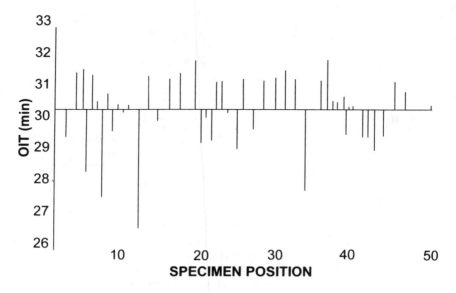

FIG. 3 -- Distribution effect of position.

TABLE 2 -- Interlaboratory test data.

ASTM Method	Temperature, °C	Pressure, MPa	Mean OIT, min	Repeatability, %	Reproducibility, %
E 1858[a]	195	0.10	29	4.3	15.
D 4565[b]	200	0.10	31	5.3	8.8
D 5885[c]	150	3.4	231	2.5	7.6
E 1858[a]	175	3.4	26	6.7	7.4
Mean				4.7	9.7

[a] Interlaboratory test performed in 1995 in support of E1858 as reported in Research Report E37-1018. Degrees of freedom df = $(n-1)(p-1)$ = 7.

[b] Obtained from an independent interlaboratory test and are not taken from the research report for this method. df = 24.

[c] df = 15.

subgroups and the subgroup means and ranges are compared to the overall mean and range. This statistical treatment found no trends or clusters either for the subsets of $n = 3$ or 4. Therefore, the material is considered to be statistically homogeneous for the OIT determination.

Mean Values

The polyolefin film sample was used in a number of interlaboratory studies (ILT) aimed at generating precision and bias statements for several different OIT ASTM standards. The results of these ILT are presented in Table 2 showing the mean OIT, within laboratory relative repeatability (= r x 100% / mean) and between laboratory relative reproducibility (= R x 100% / mean). Overall pooled relative repeatability is approximately 4.7%, and pooled relative reproducibility is 9.7% This is consistent with the interlaboratory study rule of thumb that reproducibility is anticipated to be twice repeatability.

TABLE 3 -- Effect of temperature.

Tempera-ture, °C	Oxidative Induction Time, min	
	at 0.10 MPa O_2	at 3.4 MPa O_2
200	30.0	9.8
190	---	14.4
180	96	25.2
175	132	---
170	---	47.1
165	385	---
160	676	95.4
155	882	---
150	---	208.
Pooled rel. std. dev.	4.1%	2.4%

Effect of Temperature

The OIT value is known to be strongly dependent on test temperature, with individual laboratories choosing slightly higher or lower test temperatures than the standard to meet local needs [1]. This reference material was tested in replicated ($n > 5$) determinations as a function of temperature to observe this effect both at 0.10 and 3.4 MPa oxygen

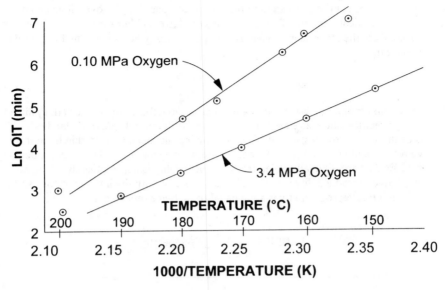

FIG. 4 -- Effect of temperature.

pressure. The results of these tests are presented in Table 3. The same data are shown in graphical form in Fig. 4 where the logarithm of the OIT value is displayed versus the reciprocal absolute temperature. The straight line plots confirms the Arrhenius dependence of the OIT value as a function of temperature and the first order form of the general rate equation governing the process.

Effect of Pressure

Some OIT experiments are carried out under elevated oxygen pressure to shorten the analysis time or lower the test temperature to avoid volatilization of the antioxidant package. Table 4 shows the effect of oxygen pressure on the proposed reference material at two commonly used test temperatures.

The same data are presented in graphical form in Fig. 5 which displays the OIT values on the ordinate and oxygen pressure on the abscissa in a log-log form. The slope of the two lines appear nearly parallel.

TABLE 4 - Effect of pressure on OIT.

Oxygen Pressure, MPa	Oxidative Induction Time, min	
	at 170°C ·	at 180°C
0.10	558.9	198.9
0.79	113.6	47.6
2.17	63.0	26.8
3.55	48.9	21.4
5.27	37.3	17.2
7.00	34.0	14.0
Pooled rel. std. dev.	7.6%	13%

Stability

The OIT value of the materials was tested using ASTM Test Method D 4565, in a single laboratory, by a large number ($n > 10$) of replicate measurements on several occasion, over a five-year period, providing an opportunity to study the long-term stability of the material. The mean values for the OIT determinations in a single laboratory over a series of nearly five years is presented in Table 5 with a pooled standard deviation for the series of measurements of ± 1.3 min. These values are statistically different from each other based on the student t tests. Taking the earliest data obtained in October 1990 to represent 100%, subsequent tests in July of 1992 and August of 1995 show a 95 and 91% relative OIT value indicating that the OIT value of the polyethylene is decreasing with time.

OIT is generally considered to follow first-order kinetics. The straight line plots of Fig. 4, for example, suggest first-order kinetics. If the data from Table 5 are best fit to the first-order kinetic expression, then the decay of the material's OIT value with time may be estimated into the future. Table 6 shows the best fit OIT values calculated from this fitted kinetic expression. The first six values represent fitted historical data points and may be compared to actual experimental results in Fig. 6. The remaining points estimate future performance. Actual historical experimental data points taken from Table 3 and 5 are plotted on the curve at the corresponding date along with their experimental error bars.

The set of data is also presented in Fig. 5 in which OIT values are plotted as function of date. The use of that Table 6 and Fig. 6 permit the individual laboratory to estimate the OIT value at times into the future. In light of the scatter in the data, such an extrapolation

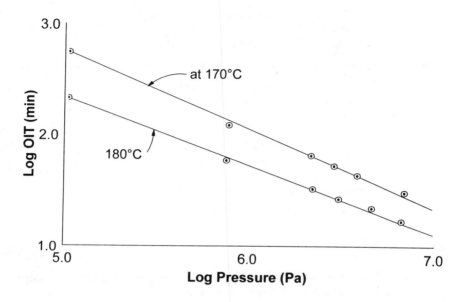

FIG. 5 -- Effect of oxygen pressure.

TABLE 5 -- Effect of time.
(temperature = 200°C; oxygen pressure = 0.10 MPa)

Date Tested, mo/yr	Mean OIT, min	Relative OIT, %
10/90	33.0	100.0
07/92	31.5	95.4
08/95	30.0	90.9

TABLE 6 -- Calculated OIT values and the effect of time.

Date, mo/yr	OIT, min
01/90	34.28
01/91	33.33
01/92	32.40
01/93	31.51
01/94	30.64
01/95	29.81
01/96	28.97
01/97	28.16
01/98	27.38
01/99	26.63
01/00	25.89
01/01	25.17
01/02	24.47
01/03	23.80
01/04	23.14

must be used with caution as ongoing efforts will be needed to verify this change of the OIT value with time.

Conclusion

In summary, a polyethylene film sample is proposed as a reference material for OIT testing. The OIT values of the material have been thoroughly tested by at least nine interlaboratory test programs over a period of more than five years, making this one of the most well-characterized OIT materials. Further, the material has been found to be statistically homogeneous, a necessary condition for service as a reference material. As a parameter dependent on test time and temperature, the OIT value appears to be decreasing with time but in a well behaved and predictable manner. The use of a table and graph permit the user of the material to estimate its OIT value in the future. The effect of temperature and oxygen pressure have also been explored permitting the user of the reference material to estimate OIT values under experimental conditions different than those of the standards. At present, this material is considered the best available reference material for oxidative induction time testing.

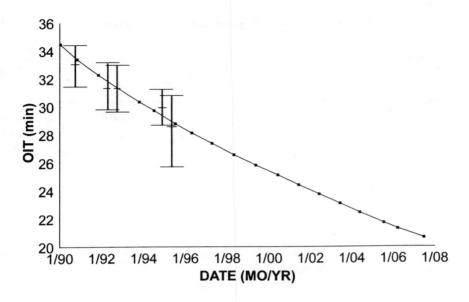

FIG. 6 -- <u>OIT decay with time.</u>

REFERENCES

[1]   Blaine, R. L., Lundgren, C. J. and Harris, M. B., "Oxidative Induction Time - A Review of DSC Experimental Effects," in this volume.

[2]   Triola, M. F., <u>Elementary Statistics</u>, Addison-Wesley Publishing Co., Reading MA, 1992, p. 84.

[3]   Mandel, J., <u>The Statistical Analysis of Experimental Data</u>, Dover Publications, New York, 1964, pp. 81-84.

Rudolf Riesen[1], Rolf Truttmann[2], and Kristy Schiano[2]

ABBREVIATED OXIDATIVE INDUCTION TIME METHOD WITHOUT HEATING/COOLING AND GAS SWITCHING

REFERENCE: Riesen, R., Truttman, R., and Schiano, K. **"Abbreviated Oxidative Induction Time Method Without Heating/Cooling and Gas Switching,"** Oxidative Behavior of Materials by Thermoanalytical Techniques, ASTM STP 1326, A. T. Riga and G. H. Patterson, Eds., American Society for Testing and Materials, 1997.

ABSTRACT: The well known standardized procedures of oxidation induction time (OIT) determination incorporate the fast heating of a specimen under nitrogen to the set temperature. After stabilization, the furnace atmosphere is changed to oxygen and from this point on the induction period is recorded. Usually, the differential scanning calorimeter (DSC) heating curves are not being used for further analysis. After the event has occurred, the cell is cooled down to ambient in order to load the next specimen. The new proposed method (SHORT) reduces the usual procedure to the necessary measurement, i.e. the analysis of the specimen under oxygen at the elevated isothermal condition. This method saves the heating and cooling period, leading to a reduction of up to 50 % of the experimental time. In addition, it is not necessary to change the atmosphere from inert to oxygen. This saves nitrogen, the investment for gas switch equipment and eliminates variations of the OIT values due to a not perfectly well defined start point, i.e. how fast the oxygen concentration is changed.

KEYWORDS: oxidation induction time (OIT), polyethylene, experimental time, isothermal method, gas switching.

[1] Manager Marketing Support Materials Characterization, Mettler-Toledo AG, Analytical, CH-8603 Schwerzenbach, Switzerland
[2] Product Management, Mettler-Toledo, Inc., Hightstown, N.J. 08520-0071, USA

INTRODUCTION

Oxidative induction time (OIT) is a relative measure of the degree of oxidative stability of the material evaluated at the isothermal temperature of the test. Existing and newly developed methods always include a cooling cycle of the furnace for loading the sample close to ambient temperatures [1 to 4]. In order to prevent oxidation already occurring during the heating phase, nitrogen purge is established. Having reached the set isothermal temperature, the purge of the furnace atmosphere is switched to pure oxygen and the oxidation induction time starts to count up.

In the standards, the measuring curve is used only for determination of the OIT value, despite the fact that the whole heating-up phase is recorded. Therefore, the valuable information of the melting behavior measured at heating rates between 20 and 40 K/min is discarded. Due to the nitrogen purge, the influence of the heating up phase on the OIT values should be negligible. The time necessary to reach the set isothermal temperature of 210 °C is approximately 8 min, including the usual waiting phases. Cooling back to the start temperature at the end of the experiment may also last several minutes and extends the total experimental time. Many differential scanning calorimeter (DSC) cells allow samples to be inserted and removed at elevated temperatures as high as 300 °C.

All this leads to the conclusion that for pure OIT determination the cooling cycle can be avoided to save experimental time, nitrogen purge and gas switch facilities.

NEW SHORT OIT METHOD

An abbreviated OIT test method is proposed which omits the cooling cycle. All other parameters and conditions are maintained as usually specified. The temperature program applied is reduced to the isothermal phase only, e.g. the DSC cell is kept all the time at the set temperature of 195 °C. The specimen is quickly inserted manually and the furnace closed before the sample reaches the set temperature (indicated by a decreasing endotherm heatflow). Automated loading of the sample is preferred because of better reproducible start conditions. Purge of the furnace atmosphere by the usual oxygen flow rate of 50 mL/min is started before the specimen reaches the set temperature. Uninterrupted purge may also be applied. After reaching the end of the measurement, the specimen is removed from the cell, which is still at the isothermal set temperature. The measuring cell is immediately ready for the next experiment.

Due to the same conditions for oxygen exposure, the same OIT values are expected.

EXPERIMENTAL PROCEDURE

Samples: To check the new SHORT method, the following samples have been used:

1) Black stabilized polyethylene (Hostalen GM5040T12) in pellet form. These pellets have been cut horizontally into two flat pieces to give approximately cylindrical shapes of 4 mm diameter and 1 mm thickness. During the analysis, the PE melted and covered the bottom of the pans (diameter 6 mm) totally.

2,3) ASTM round robin reference materials [4]: oil (C), and polyethylene (D) in form of a thin film.

Instrumentation: Mettler Toledo DSC821^e with the STAR^e software. For routine measurements, the system is fitted with sample robot and gas controller to activate the right gas flow for each segment. Manual or automated loading of the DSC cell can be performed at any temperature. The reference crucible is always of the same type as the sample crucible (40 µL aluminum without lid), but empty.

Due to short and small diameter tubings the gas switch over from nitrogen to oxygen takes place within 20 s. Using the SHORT method, the measurement starts after inserting the crucible with 35 s stabilization time for the given conditions (25 s to equilibrate the temperature and 10 s settling period).

OIT procedures: The mentioned standards [4] were used, including calibration of the measuring cell using indium and tin at a rate of 1 K/min. But, the sample masses used deviated strongly from the given standard [4]. Nevertheless, the used procedure will be abbreviated as ASTM.

The STARe system allows for automated termination of the experiment if the oxidation reaches 3 mW followed by a waiting time of 10 min to pass the exotherm maximum. Therefore, the curves are ending at different times.

Determination of the start of the oxidation (OIT): The OIT value is traditionally the time from the switch-over from nitrogen to oxygen to the extrapolated onset. In the SHORT method, the time for the OIT value starts at the begin of data collection, i.e. the start of the experiment after the above mentioned settling period. The extrapolated onset is defined by the intersection of the extrapolated baseline with the tangent to the inflection point.

All diagrams with the DSC curves display the exotherm direction upwards.

RESULTS

As a reference for the following comparisons, the measurements at 210 °C, following the standard [4], have been selected.

The following measurements show the coincidence of the two methods (Fig. 1 to 3 and Tab. 1). Hence, the OIT values determined by the new SHORT method are comparable to the standard method. The SHORT method reduces the experimental time by approximately 10 min to 48 min in total for the experiments as shown in Fig. 1.

Tab. 1: OIT values of the black PE sample measured by the ASTM method and the SHORT method at 210 °C each.

Method	Sample weight mg	OIT min
ASTM	10.7	35.3
ASTM	8.4	34.4
ASTM	12.4	35.9
prEN728 [3]	10.6	35.5
SHORT	10.9	35.7
SHORT	13.6	34.9
SHORT	9.7	35.9

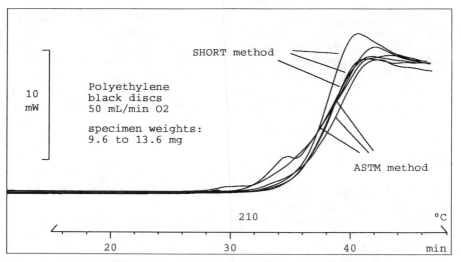

Fig. 1 — Comparisons of the DSC curves: OIT determinations of the black PE samples using the SHORT and the ASTM method respectively. The abscissa shows the elapsed time of oxygen exposure at 210 °C and gives a picture of possible variations of the curve shapes encountered.

As can be observed in the Fig. 1 to 3, the SHORT method does not only fit the OIT values of the standard method, but shows slightly better reproducibility.

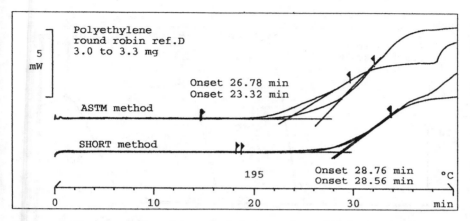

Fig. 2 — OIT measurements of the polyethylene sample used in the round robin test. Methods used: ASTM, [4], with cooling/nitrogen cycle, and the new SHORT, with only the isothermal oxygen exposure at 210 °C.

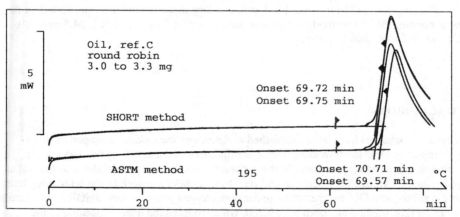

Fig. 3 — OIT measurements of the oil sample used in the round robin test. Methods used: ASTM, [4], with cooling/nitrogen cycle, and the new SHORT, with only the isothermal oxygen exposure at 210 °C. The endotherm deviation of the DSC curves in the first minutes are due to evaporations.

As already demonstrated in [5], the OIT values are strongly influenced by the selection of the isothermal temperature. Fig. 4 shows the respective curves measured using the SHORT method.

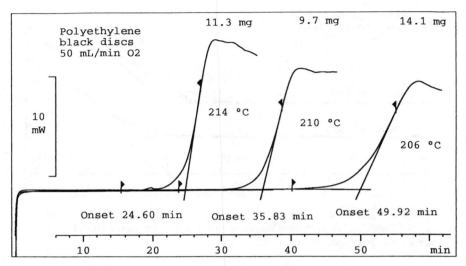

Fig. 4 — Influence of temperature on OIT values by 4 °C intervals between 206 and 214 °C, using the ASTM method; black polyethylene discs. The OIT values found by the corresponding ASTM method for the same sample are 23.8 min (214 °C), 34.5 min (210 °C) and 48.2 min (206 °C) respectively.

CONCLUSION

Standard procedures have been established to determine the oxidative properties of hydrocarbons by DSC to allow comparison of influences of the nature of the material, of additives, and of aging processes. In these standards the samples are loaded into the cell at approximately ambient temperature and then heated under nitrogen to the set temperature for the subsequent OIT measurement under pure oxygen purge. A new SHORT oxidation induction time method has been introduced, which reduces the usual procedure to the necessary analysis of the specimen under oxygen at the elevated isothermal condition. Various materials at different temperature levels have been tested for comparison of the OIT methods. The results of the new method are identical to the standard method within the normal experimental variation. The new SHORT OIT method reduces the experimental time and avoids the nitrogen purge.

REFERENCES

[1]   ASTM D3895, Standard test method for oxidative induction time of polyolefines by differential scanning calorimetry

[2]   ASTM D3350, Standard test method for polyethylene plastic pipe and fitting materials

[3]   prEN 728:1993, Plastics piping and ducting systems - Polyolefin pipes and fittings - Determination of oxidation induction time.

[4]   Project TM-01.10A of ASTM Committee E37, Standard test method for determining oxidation induction time of hydrocarbons by differential scanning calorimetry.

[5]   Truttmann, R., Schiano, K., and Riesen, R. "Oxidative Induction Time (OIT) Determinations: Influences of Temperature, Pressure, and Crucible Materials on the Result," Oxidative Behavior of Materials by Thermoanalytical Techniques, ASTM STP 1326, A .T. Riga and G. H. Patterson, Eds., American Society for Testing and Materials, 1997.

Rickey J. Seyler[1] and Roger Moody[2]

OXIDATION AS A CONTAMINATION IN MEASUREMENTS INVOLVING
SELF-GENERATED ATMOSPHERES

REFERENCE: Seyler, R. J. and Moody, R., **"Oxidation as a Contamination in Measurements Involving Self-Generated Atmospheres,"** *Oxidative Behavior of Materials by Thermal Analytical Techniques, ASTM STP 1326*, A. T. Riga and G. H. Patterson, Eds., American Society for Testing and Materials, 1997.

ABSTRACT: Recent task group activities in support of ASTM Committee E37 efforts to characterize volatile chemicals with thermal analysis have involved the use of precision pinholes in sealed specimen containers. During the laboratory testing in support of these efforts, occasional artifacts were encountered. In particular, the development of ASTM Test Method for Determination of Vapor Pressure by Thermal Analysis (E 1782) showed some sporadic exothermic activity near the onset to boiling at test pressures significantly greater than ambient. This exothermic activity is associated with partial oxidation of the specimen resulting from residual oxygen as a contaminant in the vapor, consisting otherwise of a self-generated atmosphere within the specimen container. This oxidation occurs only when the working pressure is significantly greater than ambient and the boiling temperature of the specimen approaches 200°C. The source of the oxygen contaminant is laboratory air trapped with the specimen during encapsulation and laboratory air from the DSC sample chamber partitioned into the specimen container during equilibration of the applied test pressure across the pinhole. This contamination reaction can be avoided either by initially drawing vacuum and subsequently pressurizing the sample chamber of the DSC with inert gas to the desired pressure, or by flushing the sample chamber with inert gas in a series of nominally eight successive pressurize-vent-repressurize repetitions to the desired pressure with inert gas before initiating the heating program.

KEYWORDS: oxidation, vapor pressure, pinhole, differential scanning calorimetry (DSC)

---

[1]Research Associate, Materials Science and Engineering Division, Eastman Kodak Company, Rochester, NY 14650-2158.

[2]Thermal Technologist, Imaging Research and Advanced Development, Eastman Kodak Company, Rochester, NY 14650-2132.

ASTM Subcommittee E37.01 has advanced the use of precision pinholes in sealed specimen containers as an essential ingredient for successful studies of volatile chemicals using thermal analysis. More specifically, the subcommittee has focused its attention on the measurement of boiling points and subsequent determination of vapor pressure using differential scanning calorimetry (DSC)[3] and on the determination of volatility as the mass loss per unit time using thermogravimetry (TGA).[4] For the case of boiling point determinations, the recommended 0.025 to 0.12-mm diameter pinhole in the lids of DSC specimen containers allows achievement of the equilibrium conditions necessary for boiling. This is accomplished by saturating the very small specimen container vapor space with specimen vapor while controlling leakage of vapor molecules into the sample chamber of the DSC and avoiding establishment of a significant pressure differential across the orifice. The result is generally sharp boiling endotherms with minimal preboiling vaporization.

The successful application of the pinhole to DSC studies for boiling point became the basis for the recent ASTM Test Method for Determination of Vapor Pressure by Thermal Analysis (E 1782). During development of this test method, ASTM Task Group E37.01.05 encountered several artifacts in certain DSC curves that included among others exotherms typically just before boiling.[5] These exotherms occurred only with organic samples at pressures generally above 2.0 MPa. These exotherms were assumed to be associated with degradation and should be avoided when generating vapor pressure data. Because of this and other hardware considerations, the upper pressure limit for application of ASTM Test Method E 1782 was recommended to be 2 MPa.

The generally small energy associated with these occasional exotherms relative to the heat of vaporization indicated that only a portion of the specimen was undergoing this "degradation" and suggested a contamination problem. Because high-purity solvents were being tested, and because this was not observed with water even at pressures of 2.5 MPa, it was assumed that the contaminant was introduced during handling and was likely to be oxygen.

This paper reviews a series of DSC measurements that were conducted to verify the supposition that the exotherms observed during development of ASTM Test Method E 1782 were the result of partial oxidation. Experimental conditions were identified that avoid this contaminant reaction at the upper extremes of pressures recommended for determining vapor pressure by thermal analysis.

EXPERIMENTAL PROCEDURE

DSC measurements were conducted in two laboratories using DSC

---

[3]ASTM Task Group E37.01.05 on Vapor Pressure by Thermal Methods.
[4]ASTM Task Group E37.01.19 on Volatility.
[5]Unpublished data, ASTM Task Group E37.01.05 Information Study, 1991- 1995.

instrumentation with pressure cells from TA Instruments, Inc. In Laboratory 1, a Model 2200 controller with a 910 cell base was coupled to the pressure cell. The lid of the pressure cell was modified to include a stainless steel fitting to allow coupling with an MKS Type 122A 10 000-torr (1 333-kPa) pressure transducer. Dry nitrogen or air was supplied to pressurize the cell through the purge inlet and vacuum was drawn through the purge outlet. Instrumentation in Laboratory 2 included a Model 2200 controller with a 910 cell base coupled to the pressure cell. Dry nitrogen or air was supplied to pressurize the cell through the bleed valve while the sample chamber pressure was monitored with an MKS Type 310BES 10 000-torr (1 333-kPa) pressure transducer attached to the purge inlet.

The organic samples (cyclohexane, *p*-dioxane, and toluene) in this experiment were all ACS Reagent grade. The water was distilled and deionized. Liquid specimens of 2 to 6 mg were encapsulated in hermetic DSC specimen containers the lids of which were modified with a 0.08-mm-diameter pinhole. Encapsulation was accomplished in laboratory air using either aluminum or coated aluminum specimen containers. After loading the specimen into the DSC, the sample chamber was sealed and pressurized with air or nitrogen before testing in one of three ways: the chamber was filled with the desired gas to the test pressure, the chamber was first partially evacuated and then backfilled with the desired gas to the test pressure, or the chamber was filled with the desired gas to the test pressure, vented to near ambient pressure, and refilled to the test pressure with eight repetitions. The DSC curves were recorded in general accordance with ASTM Test Method E 1782 using either a 5 or 10 °C /min heating rate.

RESULTS

Figure 1 represents typical boiling endotherms for the organic liquids of interest recorded by DSC at ambient pressure when a pinholed specimen container as recommended in ASTM Test Method E 1782 is used. However, during development of the test method by ASTM Task Group E37.01.05, the occurrence of a small exotherm at or near the onset to boiling was first reported by Schaumann[6] (Figs. 2 and 3) when test pressures in excess of 2.5 MPa were used with certain organic liquids but not with water. These small exotherms were taken to be partial degradation of the specimen and presumed to be oxygen related although measurements were recorded with a pressurized nitrogen atmosphere in the sample chambers. The lesser pressure range capability of some commercial DSCs, the expected preponderance of the use of the test method near or below ambient pressure, and the potential for avoiding such an artifact led the task group to recommend an upper test pressure limit of 2 MPa.

With further promotion of the use of the precision pinhole in DSC experiments for volatile chemicals [*1-4*],[7] it was decided that these small exotherms should be examined in

---

[6]C. Schaumann, Report to ASTM Task Group E37.01.05, April 1993.
[7]R. J. Seyler, Report to ASTM Task Group E37.01.19, September 1994.

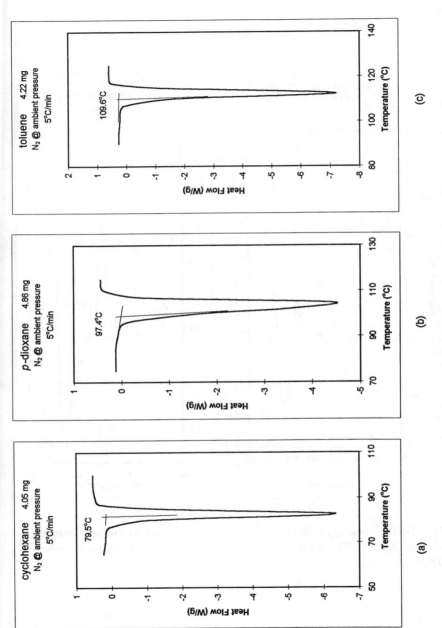

FIG. 1--*DSC boiling endotherms at ambient pressure: a*) *cyclohexane, b*) *p-dioxane, and c*) *toluene.*

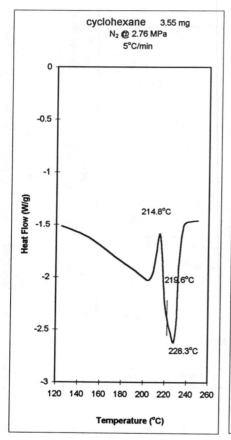

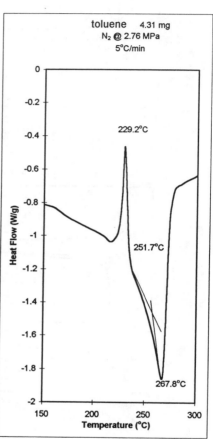

FIG. 2--*DSC curve of cyclohexane in nitrogen at 2.76 MPa with a 0.025-mm pinhole.*

FIG. 3--*DSC curve of toluene in nitrogen at 2.76 MPa with a 0.025-mm pinhole .*

more detail. The intent was to substantiate the premise that they are, in fact, oxidation, to determine their origin, and to eliminate their presence. For this purpose, four liquid materials including cyclohexane, p-dioxane, toluene, and water were examined. The four liquids were initially examined with a pressurized air atmosphere in the sample chamber of the DSC to demonstrate whether oxidation would occur and how it would appear. DSC curves were recorded at applied air pressures in the sample chamber of 0.25, 0.45, and 1.3 MPa. The results are summarized in Table 1, and representative curves illustrating the oxidation of the three organic liquids are included as Figs. 4 through 6. At the lower pressures of air, there was no evidence of any exotherm with the possible exception of cyclohexane at 0.45 MPa. At 1.3 MPa, substantial exotherms are observed before any vaporization thereby establishing the approximate conditions under which oxidation may be observed as a contaminant reaction in boiling point studies of these volatile organic liquids. As expected, there was no evidence of any exothermic activity with the water. Not only does this further substantiate the original premise that the artifacts were oxidation but it eliminates the specimen container and the pinhole as possible causes.

Having established oxidation as a possible alternative thermal event in DSC studies of volatile liquids at elevated pressures with pinhole containment, it was then necessary, using the cyclohexane, p-dioxane, and toluene, to attempt to recreate the conditions giving rise to small exotherms that were previously observed with an inert environment.

TABLE 1--*Oxidation conditions with 0.08- mm pinhole.*

Sample	T_{bp}, $°C^a$	T_{exo}, $°C^b$
Cyclohexane		
@0.25 MPa	112.2	ND^c
@0.45 MPa	135.7	126.8^d
@1.30 MPa	186.9	172.4
p-Dioxane		
@0.25 MPa	129.9	ND
@0.45 MPa	156.6	ND
@1.30 MPa	178.5	162.8
Toluene		
@0.23 MPa	140.5	ND
@0.45 MPa	172.0	ND
@1.30 MPa	>221	217.2

[a] Extrapolated onset temperature.

[b] Peak maximum temperature.

[c] ND = not detected.

[d] Possible minor exotherm.

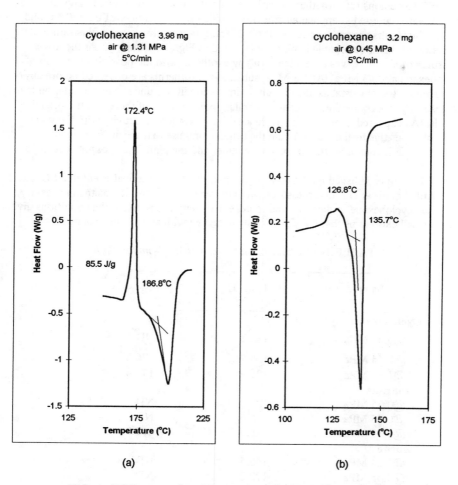

FIG. 4--*DSC curve of cyclohexane in air with a 0.08-mm pinhole:
a) at 1.31 MPa and b) at 0.45 MPa.*

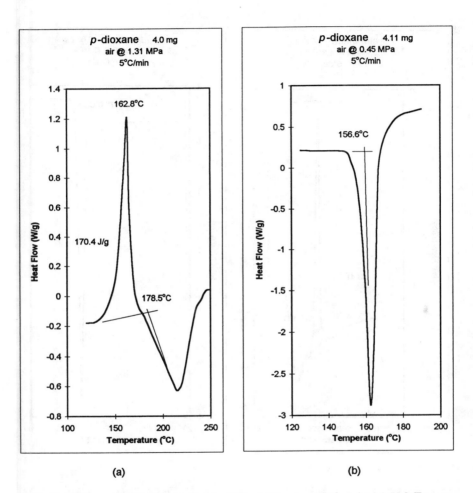

FIG. 5--*DSC curve of p-dioxane in air with a 0.08-mm pinhole: a) at 1.31 MPa and b) at 0.45 MPa.*

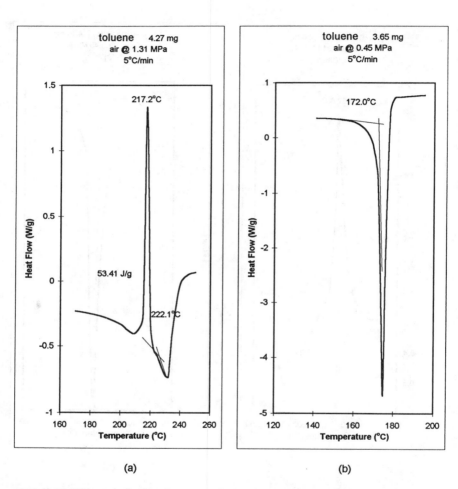

(a)   (b)

FIG. 6--*DSC curve of toluene in air with a 0.08-mm pinhole: a) at 1.31 MPa and b) at 0.45 MPa.*

Specimens of the three organics and of water were encapsulated in "laboratory air" in individual containers. After loading such an encapsulated specimen into the DSC, the sample chamber was sealed in ambient conditions and charged with pure, dry nitrogen to the desired pressure before initiating the heating program. For all three organic materials (Figs. 7 through 9), a small exotherm was observed in the DSC curves for pressures of approximately 1.4 MPa; while, as expected for water, no exotherm was observed. The peak temperatures for the exotherms are similar to those discussed above for a pressurized "air" atmosphere but significantly reduced in peak size. At a lesser pressure of approximately 1.1 MPa, there is some evidence for exothermic activity for cyclohexane and p-dioxane but not for toluene. These results again support the original premise that partial oxidation is a possible contaminant reaction when inert gas pressures significantly above ambient are used to study volatile organic materials in pinholed DSC specimen containers.

TABLE 2--*Sample chamber pressurization with nitrogen.*

Sample/Pressurization	Test Pressure, MPa	Tbp, $^{\circ}C^a$	Texo, $^{\circ}C^b$
Cyclohexane			
pressurized directly	1.45	188.7	176.6
	1.10	178.2	157.7[c]
evacuate-pressurize	...	...	...
pressurize-vent- repressurize 8 repetitions	1.38	191.7	ND[d]
p-Dioxane			
pressurized directly	1.45	172.3	164.4
	1.10	196.3	152.3[c]
evacuate-pressurize	1.38	210.4	ND
pressurize-vent- repressurize 8 repetitions	1.38	211.1	ND
Toluene			
pressurized directly	1.43	225.4	209.8
	1.10	220.8	ND
evacuate-pressurize	1.45	233.3	ND
pressurize-vent- repressurize 8 repetitions	1.44	238.1	ND

[a] Extrapolated onset temperature.

[b] Peak maximum temperature.

[c] Possible minor exotherm.

[d] ND = not detected

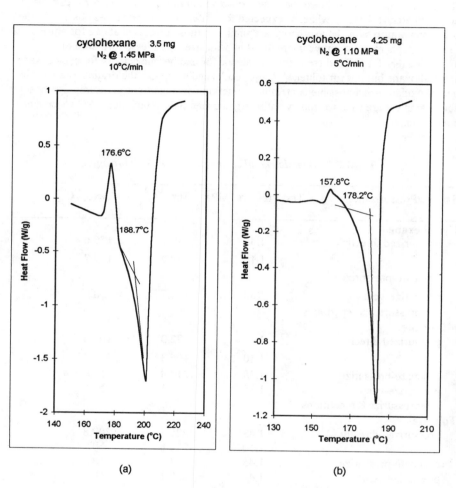

FIG. 7--*DSC curve of cyclohexane in nitrogen with a 0.08-mm pinhole: a) at 1.45 MPa and b) at 1.10 MPa*

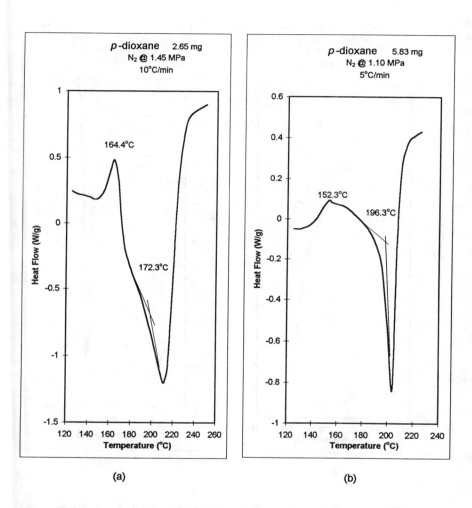

FIG. 8--*DSC curve of p-dioxane in nitrogen with a 0.08-mm pinhole: a) at 1.45 MPa and b) at 1.10 MPa.*

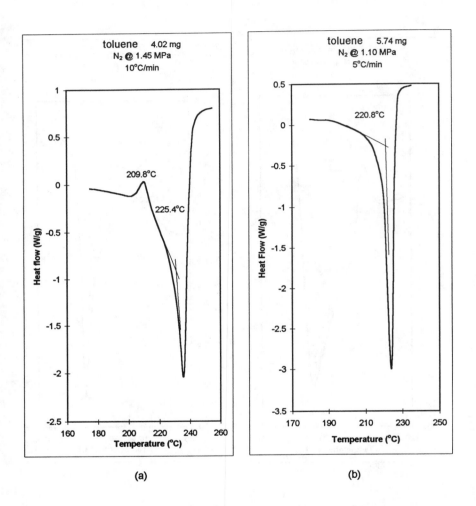

FIG. 9--*DSC curve of toluene in nitrogen with a 0.08-mm pinhole: a) at 1.43 MPa and b) at 1.10 MPa.*

Additional data summarized in Table 2 further substantiate the exothermic activity as oxygen contamination and demonstrate how it can be avoided. Whether initially evacuating the sample chamber before pressurizing with a high-purity inert gas (Fig. 10) or flushing the sample chamber by repeatedly pressurizing-venting-repressurizing eight times in succession with a high-purity inert gas like nitrogen (Fig. 11), the exotherm was avoided. This simple change of procedure by which the sample chamber is pressurized suggests the contamination source is the sealed atmosphere of the sample chamber, which like the specimen container is sealed in "laboratory air."

It was not initially obvious why the sample chamber atmosphere should be the primary source of the oxygen contamination. Although initially sealed in "laboratory air," the chamber was pressured by introducing high-purity nitrogen. In addition, a precision pinhole was included with the specimen container that was believed to serve as a relief valve that kept the specimen in and the sample chamber atmosphere out while minimizing any pressure differential across its length. As such, one would be more inclined to argue that the residual air retained within the specimen container after encapsulation would be the contaminant source. This oxygen may in fact participate in oxidation, but its concentration is believed too small to be observable in the DSC.

Rethinking the function of the pinhole with these new data helps to explain the role of the sample chamber atmosphere. Earlier work with the pinhole indicated it provided a means of establishing a self-generated atmosphere within the specimen container with minimal loss of condensed specimen. This is possible because the pinhole controls leakage of vapor molecules from the container at a rate of essentially one at a time thereby promoting rapid saturation of the specimen container vapor space. The resulting perception was that the pinhole behaved similarly to a one-way relief valve that isolated the specimen vapor space from the balance of the DSC cell. At ambient pressure, the pinhole does essentially keep the specimen in and the sample chamber atmosphere out much like a valve. However, if a momentary pressure differential of any significance is developed across the pinhole length (lid thickness), there will be an exchange of gas molecules through the pinhole to equilibrate this pressure differential. Under these conditions, the pinhole behaves more like a bidirectional restrictor pressure relief valve.

If one simply seals the sample chamber and pressurizes it, the momentary pressure differential will force gas molecules from the sample chamber into the specimen container until the pressure is equalized on both sides of the pinhole. Despite the use of an inert gas to pressurize the sample chamber that will serve to dilute the oxygen concentration, a percentage of the gas molecules entering the specimen container will be oxygen from the original "laboratory air." When added to the oxygen molecules already present from sealing the specimen container in "laboratory air," this additional oxygen may enhance oxidation to an extent of being detected if sufficient pressure is applied. The addition of nitrogen during pressurization will change the nitrogen:oxygen ratio of molecules entering the pinhole during equilibration and may actually reduce the oxygen concentration inside the specimen container. The total number of oxygen molecules available for reaction inside the specimen container will, however, have increased in proportion to the applied

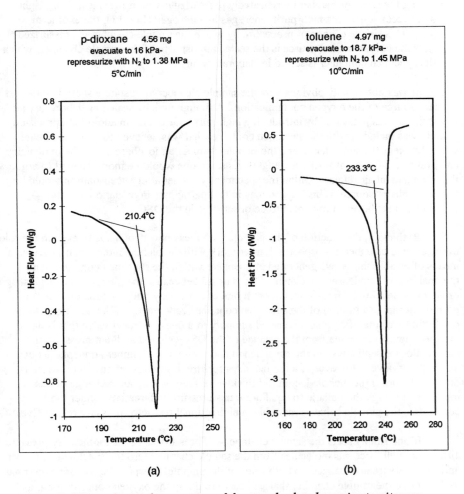

(a)   (b)

FIG. 10--*Effect of partial evacuation of the sample chamber prior to nitrogen pressurization: a) p-dioxane; 16 kPa to 1.38 MPa and b) toluene; 18.7 kPa to 1.45 MPa.*

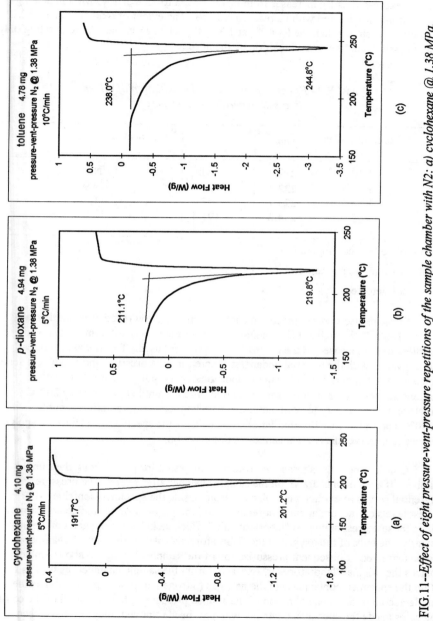

FIG. 11–*Effect of eight pressure-vent-pressure repetitions of the sample chamber with N2: a) cyclohexane @ 1.38 MPa, b) p-dioxane @ 1.38 MPz, and c) toluene @ 1.44 MPa.*

pressure. This is demonstrated in Table 3 and Figs. 9b and 12. The size of the oxidation exotherm expressed as J/g increases from 0 at 1.1 MPa to 13.8 at 1.45 MPa indicating greater extent of oxidation with increased pressure. The extent of oxidation at 1.45 MPa is, however, much less than the 53.4 J/g at 1.31 MPa when air is used demonstrating only partial oxidation and the overall dilution by the nitrogen.

TABLE 3--*Effect of pressure on the oxidation of toluene in a self-generated atmosphere.*

N2 pressure, MPa	T_{bp}, °C [a]	T_{exo}, °C [b]	peak area, J/g
1.10	217.2	ND[c]	ND
1.14	222.4	194.7	4.9
1.33	223.7	197.4	11.5
1.45	234.2	191.4	13.8

[a] Extrapolated onset temperature.

[b] Peak maximum temperature.

[c] ND = not detected.

Pressure also contributes to the contaminant oxidation reaction in another way. As noted in Tables 1 through 3, below some nominal pressure no exotherm is observed regardless of the gas used or the pressurization protocol used. This nominal "critical" pressure varies with the material under test. Increasing the pressure above ambient serves to increase the temperature of boiling proportionately (vapor pressure). Oxidation reactions generally do not occur for condensed organics until approximately 200°C. Therefore, it is necessary to both increase the boiling temperature until it is in the vicinity of 200°C and to increase the amount of oxygen inside the specimen container by applying pressure to observe this contaminant oxidation reaction.

Consider now the alternative protocols suggested for pressurizing the sample chamber. If a vacuum is drawn on the sample chamber after sealing, the pressure differential serves to extract gas molecules from inside the specimen container. For modest levels of vacuum at ambient temperature, the majority of gas molecules extracted will be those corresponding to "laboratory air," which includes the oxygen contaminant. Similarly, the bulk of the oxygen in the "laboratory air" within the sample chamber will also be extracted. Subsequent pressurization of the sample chamber to above ambient replaces the original "laboratory air" extracted from the sample chamber, as well as from within the specimen container with the new gas (nitrogen in these experiments) as the pinhole accommodates equilibration of the applied pressure. Though not all of the oxygen molecules inside the specimen container will have been removed, they will have been significantly diluted to the extent to which their interaction with the specimen is

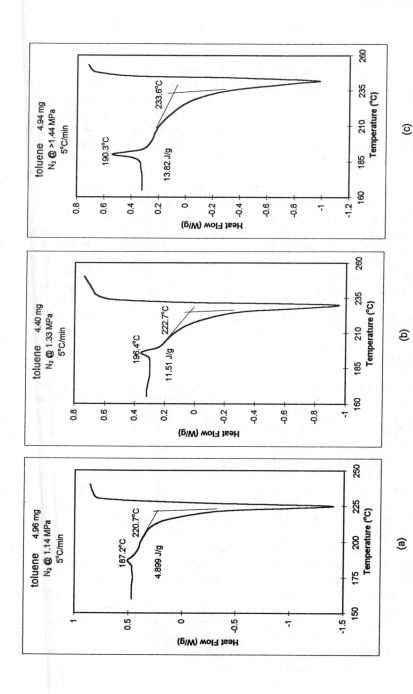

FIG. 12--*Effect of nitrogen test pressure on the oxidation contamination exotherm in toluene: a) at 1.14 MPa, b) at 1.33 MPa, and c) at >1.44 MPa.*

undetectable when an inert gas is used even at pressures sufficient to suppress boiling of the organic liquid to the vicinity of 200°C.

A word of caution may be appropriate at this point. The practitioner is reminded that if sufficient vacuum is achieved, the boiling temperature of the specimen will become ambient temperature and the actual specimen will be extracted as a gas from the specimen container. Therefore, when using this approach, it is necessary to control carefully the extent of vacuum used and, if necessary, repeat this evacuate-pressurize procedure more than once at lesser extents of vacuum if vacuum levels are limited by the specimen vapor pressure to approximately 50 kPa.

The second approach to pressurizing the sample chamber that involved eight pressurize-vent-repressurize repetitions was also effective in avoiding the contaminant oxidation exotherm. This approach "flushed" the oxygen from the sample chamber by diluting the concentration and then removing a significant fraction of the contents with each return to ambient pressure (vent step). Although the initial pressurization would have actually increased the number of oxygen molecules in the specimen container as discussed above, the venting portion creates a negative pressure differential similar to that of drawing vacuum. A portion of the specimen container vapor is thus extracted with each repetition. An increasing number of oxygen molecules are replaced by nitrogen during each successive repressurization such that after eight repetitions the number of oxygen molecules remaining is insufficient to be detected by the DSC.

CONCLUSION

DSC measurements made with cyclohexane, *p*-dioxane, and toluene encapsulated in specimen containers with 0.08-mm pinholes indicated oxidation exotherms before boiling with sample chamber atmospheres of air at pressures of 1.3 MPa or greater. When specimens of these organic materials were tested at similar conditions using a single pressurization of the sample chamber with high-purity nitrogen, small exotherms were encountered in the same temperature region just before boiling. Examining water under identical test conditions resulted in no exotherm. These results confirmed the premise established during early methods development of ASTM Test Method E 1782 that occasional small exotherms encountered during boiling point measurements at significantly elevated pressures (in excess of 1 MPa) were the result of partial oxidation. This oxidation is an undesired contamination reaction that occurs when the boiling point is elevated with pressure to the vicinity of 200°C and the amount of oxygen molecules in the specimen container is increased to a critical level. The oxygen source is the initial "lab air" that is sealed in the sample chamber of the DSC cell. Application of pressure causes oxygen molecules from the sample chamber "lab air" to be partitioned into the specimen container during equilibration across the pinhole, thereby increasing the number, though not necessarily their concentration, available for reaction with the organic specimen. Both retention of the organic specimen to temperatures in the vicinity of 200°C and increasing the number of oxygen molecules inside the specimen container are necessary to encounter this oxidation reaction in DSC studies using pinholes.

These contamination reactions can be avoided by changing the manner in which the test pressure is established in the sample chamber. Both partial evacuation with subsequent pressurization or eight pressurize-vent-repressurize repetitions with a pure inert gas have been demonstrated to reduce successfully the number of oxygen molecules inside the specimen container to a level sufficient to avoid observance of the oxidation reaction by DSC.

The description of the function of precision pinholes, such as those recommended for use in establishing self-generated atmospheres for studying volatile chemicals, has been revised from that of essentially a one-way relief valve to that of a bidirectional restrictor valve. Exchange of gas molecules occurs through the pinhole between the sample chamber and the specimen container whenever a momentary pressure differential, either positive or negative, is developed. This exchange may be used to adjust the composition of the vapor phase over the specimen at the start of the DSC heating program.

ACKNOWLEDGMENT

The authors wish to acknowledge the assistance of Dr. Sima Chervin and Mr. Paul Kingsley for their assistance in generating and preparing the pressure DSC data used in this manuscript

REFERENCES

[1] Jones, K. and Seyler, R., "Differential Scanning Calorimetry for Boiling Points and Vapor Pressure," NATAS Notes, Vol. 26, No. 2, Spring 1994.

[2] Brozena, A., et al., "Vapor Pressure Determinations Using DSC," in Proceedings of the Twenty-Second Conference of the North American Thermal Analysis Society, Sept. 1993.

[3] Casserino, M., Blevins, D., and Sanders, R., "An Improved Method for Measuring Vapor Pressure by DSC with Automated Pressure Control," in Proceedings of the Twenty-fourth Conference of the North American Thermal Analysis Society, Sept. 1995.

[4] Perrenot, B., et al., "New Pressure DSC Cell and Some Applications," Journal of Thermal Analysis, Vol. 38, 1992, pp. 381-390.

Author Index

Subject Index

A

Activation energy, 172
Additive depletion, 16
Adhesive bonding, 102
Aging, 116, 138
Aluminum, 151
Ammonium phosphate, 29
Asphalt, 138
ASTM standards, 3, 151, 184
    D 3350, 91, 151
    D 3895, 44, 91, 151
    D 4439, 76
    D 4565, 91, 151, 193
    D 5483, 91, 151
    E 1782, 212
    E 1858, 151, 164

C

Cannon, 128
Carbon fibers, 128
Carbonyl, 138
Cellulose degradation, 29
Color formation, degradation, 44
Combustion, cellulose, 29
Composites, fiber-reinforced
    epoxy, 128
Contamination, oxygen, 212
Copper
    silver-plated, 116
    tin-plated, 116
Crucible material, influence on
    oxidative induction time,
    184
Crystallinity, degree of, 102

D

Degradation, 58
    cellulose, 29
    degree, by color formation, 44
    insulation, 116
    precursor, OIT value
        reduction, 76
    thermal oxidative, 116

Differential scanning calorimetry,
    3, 16, 212
    D 3895, 44, 91, 151
    D 4565, 91, 151, 193
    D 5483, 91, 151
    heat flux, 164
    polyethylene, 76, 164, 193
    polyolefins, 58
    power compensation, 164
    pressure, 91, 151, 164, 172

E

Electrical insulation, 116
Epoxy, 128
Ethylene tetrafluoroethylene,
    116

F

Fiberite, 128
Flow rate, 3
Fourier transform infrared
    analysis, 16, 91

G

Gas, inert, chamber
    pressurization, 212
Gas mass spectrometry, 116
Gas phase modification, 102
Gas switching, 205
Geomembrane, 76
Geosynthetics, terminology
    D 4439, 76
Glass fibers, 128
Glass transition temperature, 128

H

Hydrocarbons, 151, 172

I

Induction time, oxidative (See
    Oxidative induction time)
Insulation, electrical, 116

235